SpringerBriefs in Architectural Design and Technology

Series Editor

Thomas Schröpfer, Architecture and Sustainable Design, Singapore University of Technology and Design, Singapore, Singapore

Indexed by SCOPUS

Understanding the complex relationship between design and technology is increasingly critical to the field of Architecture. The *Springer Briefs in Architectural Design and Technology* series provides accessible and comprehensive guides for all aspects of current architectural design relating to advances in technology including material science, material technology, structure and form, environmental strategies, building performance and energy, computer simulation and modeling, digital fabrication, and advanced building processes. The series features leading international experts from academia and practice who provide in-depth knowledge on all aspects of integrating architectural design with technical and environmental building solutions towards the challenges of a better world. Provocative and inspirational, each volume in the Series aims to stimulate theoretical and creative advances and question the outcome of technical innovations as well as the far-reaching social, cultural, and environmental challenges that present themselves to architectural design today. Each brief asks why things are as they are, traces the latest trends and provides penetrating, insightful and in-depth views of current topics of architectural design. *Springer Briefs in Architectural Design and Technology* provides must-have, cutting-edge content that becomes an essential reference for academics, practitioners, and students of Architecture worldwide.

Luca Salvati · Ioannis Konaxis

Vacant Land

Urban Design, Economic Expansion, and Environmental Consequences

Springer

Luca Salvati
Department of Methods and Models for Economics, Territory and Finance (MEMOTEF), Faculty of Economics
Sapienza University of Rome
Rome, Italy

Ioannis Konaxis
Department of Tourism Studies
University of Piraeus
Piraeus, Greece

ISSN 2199-580X ISSN 2199-5818 (electronic)
SpringerBriefs in Architectural Design and Technology
ISBN 978-981-95-4203-1 ISBN 978-981-95-4204-8 (eBook)
https://doi.org/10.1007/978-981-95-4204-8

This Springer imprint is published by the registered company Springer Nature Singapore Pte Ltd.
The registered company address is: 152 Beach Road, #21-01/04 Gateway East, Singapore 189721, Singapore

Contents

1 Vacant Land, from Urban Design to Environmental Decline 1
1.1 Introduction 1
1.2 Defining, Delineating, Envisaging Urban Voids 3
1.2.1 Interpreting Voids as Relational Spaces 4
1.3 Moving Across Scales, from Places to Landscapes 6
1.3.1 From Landscapes to Regions: Voids as Basic Elements of Metropolitan Areas 6
1.4 Dispersed Settlement Expansion and the Consolidation of Vacant Land 8
1.4.1 Vacant Land and Social Dynamics 8
1.5 Looking Beyond the Fringe: Abandoned Soils and Relict Landscapes 9
1.6 Concluding Remarks 11
References 11

2 Settlement Expansion and the Formation of Urban Voids 17
2.1 Introduction 18
2.2 The Spatial Dimension of Urban Voids 19
2.2.1 Toward an 'Architecture' of Urban Voids? 21
2.3 Urban Voids, Compact Cities, and Public Space 23
2.3.1 Vacant Land as a Relational Linkage Between Architecture and Nature 25
2.4 (Re)connecting Urban Design with the Ecological Functions of Vacant Land 26
2.4.1 The Role of Urban Greening 28
2.5 Concluding Remarks 30
References 30

3 Vacant Land and Urbanization Processes 35
3.1 Introduction 36
3.2 Urban–Rural Polarization and Metropolitan Growth 37
3.2.1 The Long-Term Evolution of Metropolitan Regions 39
3.3 Understanding Leapfrog Urbanization Along Metropolitan Gradients 42
3.3.1 Urban Morphology, Fringe Development and Vacant Land 44
3.4 Settlement Diffusion, Economic Functions and Urban Voids 46
3.4.1 Sub-Central Expansion and Landscape Transformations 47
3.5 Vacant Landscapes In-Between Settlement Morphology and Metropolitan Structure 49
3.5.1 Urban Growth, Vacant Land and Settlement Aestethic 50
3.6 Peri-Urban Expansion, Spatial Planning, and Vacant Land 52
3.6.1 Planning New Forms of Contemporary Cities Under a 'Land Take' Era 54
3.7 Concluding Remarks 56
References 58

4 Local Development, Social Relations, and the 'Informal' Consolidation of Urban Voids 65
4.1 Introduction 65
4.2 Culture or Economy: The Basic Ground of Cities' Development 67
4.3 The Long-Lasting Linkage Between Urban Ideology and Fringe Land 67
4.4 Concluding Remarks 68
References 69

5 Fallow Land and 'Vacant' Peri-Urban Spaces 73
5.1 Introduction 74
5.2 Fallow Land and Peri-Urban Farming 74
5.2.1 Fallow Land and Economic Downturns 76
5.3 Sustainable Soil Management, Spatial Planning and Vacant (Fallow) Land 77
5.3.1 Fallow Land and Social Dynamics 79
5.4 Concluding Remarks 81
References 81

Chapter 1
Vacant Land, from Urban Design to Environmental Decline

Abstract The present contribution examines the interplay between settlement expansion and metropolitan landscapes, focusing on the emergence and significance of urban voids at both local and regional scale. Adopting a transdisciplinary approach, it integrates architectural, geographical, social, and environmental perspectives to analyze the evolution of dispersed urban forms, peri-urban territories, and the spatial distribution of vacant land. Urban voids—intended as areas left vacant by obsolescence, abandonment, or shifts in final use—are interpreted as relational spaces with morphological, functional, social, and ecological value, serving as connective tissue within cities. This work proposes multi-criteria analytical frameworks, operational tools and visual experiences, for the sustainable management and valorization of urban voids, emphasizing the potential of these spaces to support social life, environmental restoration, and cohesive metropolitan landscapes. It argues for landscape-sensitive, interdisciplinary planning strategies that reconceptualize vacant land from residual byproduct of urban growth into multifunctional, resilient, and socially inclusive assets fostering sustainable metropolitan development.

Keywords Fringe · Land · Architecture · Settlement · Sustainability

1.1 Introduction

A complete understanding of the inherent complexity in urban–rural relationships within metropolitan regions benefits from a transdisciplinary perspective that integrates architectural knowledge with geographical, economic, social, and environmental dimensions of a given landscape (e.g. [1–3]). Building on this premise, the present study aims to represent the intricate geography of these relationships, proposing original frameworks for analysis, reflection, and design across diverse spatial scales [4], and specifically focusing on urban and peri-urban vacant land.

Considering both compact and dispersed settlement morphologies, urban voids can be envisaged as the product—sometimes unintended, sometimes deliberately created—of continuous settlement expansion governed by planning rules that are

L. Salvati and I. Konaxis, *Vacant Land*,
SpringerBriefs in Architectural Design and Technology,
https://doi.org/10.1007/978-981-95-4204-8_1

not always systemic but often fragmented, isolated and partly ineffective [5–7]. Examining metropolitan contexts in advanced economies also provides a representative sample of social dynamics, and the dominant industrial path, since World War II. This is particularly evident in regions where traditionally compact forms have progressively becoming dispersed, leaving voids of varying dimensions that require planning, design, active (economic and ecological) use, and preservation [8–10].

The morphological and functional interpretation of urban voids initially draws upon the most consolidated forms of 'western' cities, while extending its analytical horizon to more evolved metropolitan morphologies in advanced economies [11–13]. Urban and metropolitan landscapes are therefore examined and interpreted through multi-criteria frameworks, adopting a diachronic perspective to delineate processes of change over time [14–16]. Within this context, urban growth—and the consequent formation (or consolidation) of vacant land—is reconsidered through the lens of the incipient tension between compactness and dispersion (Fig. 1.1) that characterizes contemporary metropolitan environments [17–19].

From this standpoint, comparative analyses of urban voids at local scales offer conceptual and operational insights, integrating architectural design and spatial planning with a systemic understanding of peri-urban landscapes [20–22]. This approach underscores the preliminary observation that metropolitan landscapes constitute the

Fig. 1.1 The micro-scale of urban voids within residential (compact) settlements. *Source* authors' photographic archive

most complex and evolving element within a 'westernized' world—namely European and North American metropolitan regions [23–25]. Urban voids are thus interpreted as a form of landscape that reflects processes of settlement expansion, while lacking a structured order characteristic of compact, radiocentric growth driven by urban–rural income differentials [6, 14, 26]. Urban voids can be also regarded as spaces with significant potential for economic and social development, provided that negative ecological implications are carefully considered [27–29]. These voids evolve alongside the cities that contain them, integrating into peri-urban contexts with pronounced contrasts—a relevant challenge for interpretation and management [30].

1.2 Defining, Delineating, Envisaging Urban Voids

The morphological and functional interpretation of urban voids proposed in this work was based on a visual investigation of metropolitan landscapes adopting multi-criteria frameworks, possibly in a diachronic perspective, as initially delineated in Chap. 2. Sprawl, a protagonist of this investigation, is interpreted in light of the interplay between compactness and dispersion typical of contemporary cities. Starting from this assumption, the comparative analysis of urban voids—basically carried out at a local scale—proposes concepts and operational definitions, integrating architectural knowledge and spatial planning with a systemic vision of peri-urban landscapes in both European and North American contexts, as better conceptualized in Chap. 3. Voids evolve—like the cities that contain them—towards peri-urban contexts rich in gradients that are particularly difficult to interpret, as directly considered in Chap. 4.

This work highlights how urban voids do not only include the areas consolidated over time by the expansion of the compact city, but also peripheral, abandoned, and cultivated plots, often without a specific intended use, and thus including land unsuitable for urbanization. Their evaluation—from both a typological and functional point of view—is an important result for setting policy interventions aimed at the sustainable recovery of spaces, as argued in Chap. 5. The identification of these new empty spaces 'without identity and use', may contribute to inform policies of settlement containment at both local and metropolitan scales. In this sense, urban voids are identified and defined according to an original and—at the same time—simplified interpretative grid, based on documentary, bibliographic, photographic, and planning supports. These tools provide an original view of a central theme in landscape analysis, in turn stimulating a set of reflections on the protection and enhancement of areas that are surely not marginal, but whose use tends to be fragmented and spatially disjointed.

1.2.1 Interpreting Voids as Relational Spaces

More specifically, Chap. 2 examines the concept of urban voids in light of metropolitan expansion. Urban voids are understood as areas rendered vacant due to processes of obsolescence, abandonment, or changes in use, encompassing derelict land, large unbuilt spaces, and interstitial sites [31–33]. Sometimes squares, streets, and infrastructure are included in this definition. From both philosophical and architectural perspectives, voids are considered relational spaces endowed with topological, perceptual, and symbolic significance (Fig. 1.2), in which openness is created to accommodate human dwellings. In this view, place emerges through human interactions with space, transforming voids from passive gaps into relational environments [34].

This 'relational' interpretation positions voids as scalar transitions and connective tissue within metropolitan settings, supporting social life and fostering cohesion across the city (e.g. [35–37]). Historically, urban squares exemplify the evolution of voids, reflecting changes in both functions and symbolic meaning [38–40]. During the Renaissance, squares embodied civic ideals, whereas in later periods they were often designed to prioritise scenographic effect or vehicular traffic flow, thereby diminishing their role as spaces for social aggregation [41]. In contemporary contexts, squares are frequently shaped by commercial activity, residential functions, and

Fig. 1.2 Derelict land along the fringe of traditional rural settlements. *Source* authors' photographic archive

monumentality, as exemplified by market squares, gardens within residential neighborhoods, and celebratory openings centered around monuments or public buildings. Streets potentially constitute another type of void (Fig. 1.3), functioning as dynamic spaces that mediate movement, perception, and urban representation [13, 42, 43].

The phenomenon of 'pedestrianisation' beginning in the 1960s, emerged as a first response to increasing traffic congestion [19, 44, 45]. The chapter further explores vacant lots as relational interfaces between architecture and natural elements [6]. In this perspective, voids are better conceived as 'multifunctional public spaces', where the integration of natural components fosters social interaction, ecological restoration, and aesthetic experiences [46]. The deliberate incorporation of greenery, water

Fig. 1.3 Urban street trees and micro-greening design. *Source* authors' photographic archive

features, and playful architectural interventions strengthens these spaces' identity and enhances their relational capacity within the landscape (e.g. [16, 27, 47]).

1.3 Moving Across Scales, from Places to Landscapes

The relationship between cities and the surrounding countryside was examined adopting theoretical concepts and operational definitions such as 'peri-urbanisation' and 'rurbanisation', which illustrate how settlement expansion increasingly blurs the distinction between urban and rural environments [2, 48, 49]. Urbanisation projects city lifestyles into rural areas, while rural landscapes simultaneously absorb settlement morphologies, resulting in complex fringe zones characterised by hybrid functions and fluid identities [23, 50, 51]. Metropolitan growth produces fragmented urbanisation patterns, incorporating villages, industrial sites, and commercial centres into dispersed landscapes [11–13].

In this view, we addressed the role of voids (taken as a sink of natural elements) in shaping metropolitan landscapes [20]. Elements such as air, water, soil, and vegetation constitute essential ecological foundations, while anthropogenic interventions—including infrastructure and buildings—continuously transform these environments [15]. Intended as a basic form of void, urban greenery performs critical ecological, health-related, protective, social, cultural, and aesthetic functions, contributing to microclimate regulation, human wellbeing, and social cohesion [43]. Parks and gardens emerge as contemporary, collective urban spaces enabling recreation, socialisation, and ecological integration [6]. The study of parks and gardens as functional voids demonstrates how thematic design, context-sensitive integration, functional diversity, mobility systems, and natural components collectively generate successful public spaces that bridge architectural and ecological concerns (e.g. [52]. Urban voids should not be perceived merely as vacant lands, rather, they represent complex relational spaces with intertwined functional, social, aesthetic, and ecological dimensions.

1.3.1 From Landscapes to Regions: Voids as Basic Elements of Metropolitan Areas

In this perspective, Chap. 3 analyses the transformation of urban and peri-urban landscapes, focusing on the complex interactions between built and vacant land and the socio-spatial consequences of these processes [1, 9, 53]. We assume that fragmented and low-density developments—especially in suburban and fringe territories—contribute to spatially uncoordinated urban growth, challenging traditional planning paradigms. Urban diffusion is treated as both a historical and contemporary phenomenon, being observed in different forms in advanced economies, and facing kaleidoscopic, differential outcomes across European and North American cities [8, 10, 14].

A recurring emphasis is placed on how settlement sprawl often unfolds at the expense of rural and agricultural land, leading to the progressive degradation of natural landscapes, the fragmentation of ecosystems, and the emergence of 'non-places'—spaces lacking identity and social cohesion [54]. Sprawl does not correspond to a single morphology or context, but rather takes diverse forms influenced by regional planning, market dynamics, infrastructural development, and the physical geography of territories. A central focus of the discussion is the peri-urban zone, conceptualised as the transitional belt between urban and rural territories [35, 36, 50]. Initially regarded as a simple buffer, this zone has progressively become home to mixed-use, low-density developments that frequently lack integration with the broader urban and rural systems [11, 12, 55].

The contribution therefore argues for a redefinition of the fringe—not as a rigid border, but as a potential mediator between city and the countryside, capable of sustaining environmental sustainability, accommodating new economic functions such as agri-tourism and organic farming, and supporting social inclusion [56]. Within this framework, vacant land is reconceptualised not as residual space but as a significant structural element of the dispersed city [22]. These voids hold the potential to foster redevelopment, ecological restoration, and the redefinition of urban form through landscape-oriented planning approaches (Fig. 1.4).

Fig. 1.4 The visual complexity of fringe vacant land. *Source* authors' photographic archive

1.4 Dispersed Settlement Expansion and the Consolidation of Vacant Land

Low settlement density, discontinuous building patterns, decentralization, weak clustering, and limited functional integration are inherent features of sprawled landscapes [3]. These characteristics contribute to car dependency, social segregation, and inefficient land use [57–59]. In a line of continuity with previous thinking, we further discussed and criticized the aesthetic and functional disjunctions of sprawl—where malls, gated communities, and industrial zones form disconnected 'development islands', frequently replacing traditional public spaces with pseudo-public ones [23, 37, 51], and thus creating another, completely different, form of vacant land.

In discussing transport and mobility, we highlighted the limitations of car-based planning and the underperformance of public transit in sprawled areas, indirectly shaping the consolidation of an unstructured matrix of vacant land. Policy tools such as growth boundaries (e.g. greenbelts, or green–blue infrastructures), clustering, and agricultural zoning, were considered as strategies to contain sprawl and preserve open spaces, possibly consolidating voids and fringe vacant land [17]. Ultimately, the contribution calls for a systemic, interdisciplinary, and landscape-sensitive approach to spatial planning [25]. Sprawl should be addressed not only as a spatial issue, but also as a socioeconomic and environmental challenge [18]. New planning paradigms should reconcile the hybrid, multipolar, and evolving nature of contemporary cities with sustainable and cohesive development goals [60].

Moving to even broader observation scales, Chap. 4 examines social and demographic dynamics shaping metropolitan development in disadvantaged contexts of advanced countries, emphasising informal land allocation and working-class adaptation as latent factors underlying formation (and consolidation) of urban voids [61–63]. Unlike in Northern Europe or the United States, where high-income groups often relocate to suburban districts, Mediterranean cities exhibit a concentration of wealthy classes in city centres, while working classes inhabit peripheral informal settlements [64–66]. This divergence is explained through cultural preferences, commuting costs, and the persistence of informal economies [67–69].

1.4.1 Vacant Land and Social Dynamics

In Southern Europe, working-class settlements emerge from three main traits: (i) adaptation within dominant land allocation systems, (ii) creativity in establishing subordinate production modes, and (iii) 'spontaneism' expressed through informal settlements [33]. Examples include, among others, 'gecekondus' in Turkey and 'borgate' in Italy, reflecting a culture tolerant of illegality due to weak planning, social necessity, and economic marginalisation [24, 37, 65]. Informal economies involving artisans, family entrepreneurs, and free labourers coexisted with formal sectors, sustaining livelihoods and shaping metropolitan forms, giving indirect fuel to voids' consolidation [70].

Mediterranean informality often functions as a vehicle for social promotion rather than mere survival [63]. The study underscores the inadequacy of theories based exclusively on economic modes of production, instead emphasising social formations—the complex articulation of economic, political, cultural, and material forces—as central to understanding urban patterns at the base of vacant lands' spread [71]. Human ecology is presented as an analytical framework linking social classes and space, although its formal ecological models insufficiently explain Mediterranean realities [35]. Rather, urbanisation without industrialisation characterises the region, where industrial growth followed—rather than triggered—city expansion, in contrast to Northern European experiences [72–74].

1.5 Looking Beyond the Fringe: Abandoned Soils and Relict Landscapes

Moving our focus on fringe (vacant) land, the book finally addresses peri-urban landscape transformations caused by suburban sprawl, farm abandonment, and environmental degradation. It suggests integrating environmental sustainability with social and economic policies to manage fringe landscapes. The concluding argument calls for multi-scale, holistic planning approaches that recognise social agency, informal economies, and cultural traditions to create sustainable (and socially cohesive) urban futures in metropolitan contexts. More specifically, Chap. 5 examines fallow land and vacant peri-urban spaces highlighting how urban expansion into rural land may transform traditional fringes. These landscapes may have high cultural, ecological, and culinary significance, yet sprawl threatens their continuity and value, possibly fueling the spread of fallow land [20, 75, 76].

In peri-urban contexts, fertile land often becomes fallow due to real estate speculation or shifts in agricultural intensity [6, 54, 77]. The contribution assumes fallow land as a proxy of land-use intensity and fringe economic patterns. During economic growth, fallow land may increase as farmers speculated on potential development, while recession may flatten (or even reverse) this trend, with land returning to cultivation, due to reduced building activity and speculation (Fig. 1.5). Additionally, it is hypothesized how fallow land under economic subsidies and incentives (e.g. those erogated from Common Agricultural Policy in Europe) concentrated in areas with higher urban development pressure, while agricultural-specialized districts saw a generalized decrease in fallow land extension. Factors such as flexible zoning rules, ineffective planning, and the legalisation of informal settlements may additionally fuel this phenomenon [32, 67, 78].

Fig. 1.5 Relict farmland in tourism-specialized settlements. *Source* authors' photographic archive

The study highlights more recent (and broader) socioeconomic dynamics, arguing that a (more or less ample) segment of urban populations, particularly during a recessionary stage, may express interest in returning to agriculture for livelihood security. During persistent stagnation, peri-urban districts often provide mixed income opportunities, with residents engaging in multiple jobs beyond farming. Landscape transformations also reflect wider urbanisation trends, characterised by shifts from compact to dispersed urban forms, increasing land consumption and ecological fragmentation (e.g. [6, 14, 26]). In this perspective, fringe areas become mosaics of intensive agriculture, abandoned cropland, and speculative fallow land, with consequences for ecological stability and rural identity [11–13]. The work concludes that fallow land is a sensitive indicator of the interplay between economy, society, and environment. It represents both ecological and economic potential, depending on policy direction [42]. Based on these premises, effective land management strategies should integrate fallow land into broader sustainability frameworks, supporting agroforestry systems, biodiversity conservation, and cultural traditions [6, 54, 77]. Without careful governance, fallow land seems to be increasingly exposed to economic and ecological risk of degradation—undermining ecosystem services and resilience, and easily becoming a dominant component of peri-urban voids.

1.6 Concluding Remarks

The evolution of vacant land in contemporary cities reflects complex interactions between urban design, economic growth, and environmental externalities. As urban expansion continues unrested, vacant land emerges both as a byproduct of fragmented (and, possibly, spatially uncoordinated) planning, and as a latent asset within speculative urban economies. Urban design often neglects these voids, treating them as residual spaces rather than opportunities for ecological integration, public life, or landscape connectivity. Economically, vacant land embodies dual potential: during expansions, it becomes an object of speculation, withheld from productive use in anticipation of a sort of 're-zoning' and higher returns; under recession, it risks abandonment and degradation—reducing land value and social utility. Simultaneously, the environmental externalities of vacant land are significant.

Left unmanaged, such areas can fragment ecosystems, foster invasive species, and contribute to urban heat islands and soil erosion. Conversely, strategic reintegration of vacant land within urban design frameworks can mitigate these effects, supporting biodiversity, microclimate regulation, and public health. Ultimately, the evolution of vacant land should shift from unplanned byproduct to deliberate resource. Integrated approaches combining landscape-sensitive design, socially inclusive economic strategies, and ecological stewardship can transform vacant land from liability into resilient, multifunctional assets, supporting sustainable growth and urban wellbeing in an era marked by rapid demographic and economic transitions.

References

1. Brueckner JK, Fansler D (1983) The economics of urban sprawl: theory and evidence on the spatial sizes of cities. Rev Econ Stat 55:479–482
2. Couch C, Petschel-Held G, Leontidou L (2007) Urban sprawl in Europe: landscapes, land-use change and policy. Blackwell, Oxford
3. Ewing R, Pendall R, Chen D (2002) Measuring sprawl and its impact. Tech rep, Smart Growth America, Washington DC
4. Scott AJ (2001) Globalization and the rise of city-regions. Eur Plan Stud 9:813–826
5. Coluzzi R, Bianchini L, Egidi G, Cudlin P, Imbrenda V, Salvati L, Lanfredi M (2022) Density matters? settlement expansion and land degradation in peri-urban and rural districts of Italy. Environ Impact Assess Rev 92:106703
6. Salvati L (2013) Monitoring high-quality soil consumption driven by urban pressure in a growing city (Rome, Italy). Cities 31:349–356
7. Salvati L, Zitti M (2008) Regional convergence of environmental variables: empirical evidence from land degradation. Ecol Econ 68:162–168
8. Schneider A, Woodcock CE (2008) Compact, dispersed, fragmented, extensive? a comparison of urban growth in twenty-five global cities using remotely sensed data, pattern metrics and census information. Urban Stud 45:659–692
9. Tsai YH (2005) Quantifying urban form: compactness versus sprawl. Urban Stud 42:141–161
10. Turok I, Mykhnenko V (2007) The trajectories of European cities, 1960–2005. Cities 24:165–182
11. Catalán B, Saurí D, Serra P (2008) Urban sprawl in the Mediterranean? patterns of growth and change in the Barcelona metropolitan region 1993–2000. Landsc Urban Plan 85:174–184

12. Domènech E, Saurí D (2006) Urbanisation and water consumption: influencing factors in the metropolitan region of Barcelona. Urban Stud 43:1605–1623
13. Frenkel A, Ashkenazi M (2008) Measuring urban sprawl: how can we deal with it? Environ Plan B Plan Des 35:56–79
14. Egidi G, Salvati L, Vinci S (2020) The long way to Tipperary: city size and worldwide urban population trends, 1950–2030. Sustain Cities Soc 60:102148
15. Salvati L, Perini L, Sabbi A, Bajocco S (2012) Climate aridity and land use changes: a regional-scale analysis. Geogr Res 50:193–203
16. Salvati L, Ranalli F, Carlucci M, Ippolito A, Ferrara A, Corona P (2017) Forest and the city: a multivariate analysis of peri-urban forest land cover patterns in 283 European metropolitan areas. Ecol Indic 73:369–377
17. Bruegmann R (2005) Sprawl: a compact history. Univ Chicago Press, Chicago
18. Burchell RW, Shad NA, Listokin D, Phillips H, Downs A, Seskin S, Davis JS, Moore T, Helton D, Gall M (1998) The costs of sprawl—revisited. Transit Coop Res Program Rep 39. Natl Acad Press, Washington DC
19. Tsilimpounidi M (2012) Athens 2012. City 16:546–556
20. Colantoni A, Ferrara C, Perini L, Salvati L (2015) Assessing trends in climate aridity and vulnerability to soil degradation in Italy. Ecol Indic 48:599–604
21. Imbrenda V, Quaranta G, Salvia R, Egidi G, Salvati L, Prokopovà M, Coluzzi R, Lanfredi M (2021) Land degradation and metropolitan expansion in a peri-urban environment. Geomatics Nat Hazards Risk 12:1797–1818
22. Salvati L, Venanzoni G, Serra P, Carlucci M (2016) Scattered or polycentric? Untangling urban growth in three Southern European metropolitan regions through exploratory spatial data analysis. Ann Reg Sci 57:1–29
23. Gospodini A (2006) Portraying, classifying and understanding the emerging landscapes in the post-industrial city. Cities 23:311–330
24. Keyder C (2005) Globalization and social exclusion in Istanbul. Int J Urban Reg Res 29:124–134
25. McNeill D (1999) Globalization and the European city. Cities 16:143–148
26. Salvati L, Carlucci M (2011) The economic and environmental performances of rural districts in Italy: are competitiveness and sustainability compatible targets? Ecol Econ 70:2446–2453
27. Ferrara A, Kosmas C, Salvati L, Padula A, Mancino G, Nolè A (2020) Updating the MEDALUS-ESA framework for worldwide land degradation and desertification assessment. Land Degrad Dev 31:1593–1607
28. Ferrara C, Carlucci M, Grigoriadis E, Corona P, Salvati L (2017) A comprehensive insight into the geography of forest cover in Italy: exploring the importance of socioeconomic local contexts. For Policy Econ 75:12–22
29. Wilson GA, Kelly CL, Briassoulis H, Ferrara A, Quaranta G, Salvia R, Detsis V, Curfs M, Cerda A, El-Aich A, Liu H, Kosmas C, Alados CL, Imeson A, Landgrebe-Trinkunaite R, Salvati L, Naumann S, Danwen H, Iosifides T, Kizos T, Mancino G, Nolè A, Jiang M, Zhang P (2017) Social memory and the resilience of communities affected by land degradation. Land Degrad Dev 28:383–400
30. Salvati L, Mavrakis A, Colantoni A, Mancino G, Ferrara A (2015) Complex adaptive systems, soil degradation and land sensitivity to desertification: a multivariate assessment of Italian agro-forest landscape. Sci Total Environ 521:235–245
31. Economidou E (1993) The Attic landscape throughout the centuries and its human degradation. Landsc Urban Plan 24:33–37
32. Economou D (1997) The planning system and rural land use control in Greece: a European perspective. Eur Plan Stud 5:461–476
33. Kourliouros E (1997) Planning industrial location in Greater Athens: the interaction between deindustrialization and anti-industrialism during the 1980s. Eur Plan Stud 5:435–460
34. Minca C (2003) Critical peripheries. Environ Plan D Soc Space 21:160–168
35. Leontidou L (1990) The Mediterranean city in transition. Cambridge Univ Press, Cambridge

36. Leontidou L (1993) Postmodernism and the city: Mediterranean versions. Urban Stud 30:949–965
37. Leontidou L (1996) Alternatives to modernism in (Southern) urban theory: exploring in-between spaces. Int J Urban Reg Res 20:180–197
38. Allegretti G, Cellamare C (2009) The ambiguous renaissance of Rome. In: Porter L, Shaw K (Eds) Whose urban renaissance? Routledge, London
39. Krumholz N (1992) Roman impressions: contemporary city planning and housing in Rome. Landsc Urban Plan 22:107–114
40. Mudu P (2006) Patterns of segregation in contemporary Rome. Urban Geogr 27:422–440
41. Gospodini A (2001) Urban waterfront redevelopment in Greek cities. Cities 18:285–295
42. Orenstein DE, Frenkel A, Jahshan F (2014) Methodology matters: measuring urban spatial development using alternative methods. Environ Plan B Plan Des 41:3–23
43. Pacione M (2003) Urban environmental quality and human wellbeing—a social geographical perspective. Landsc Urban Plan 65:19–30
44. Phelps A, Parsons A, Ballas D, Dowling A (2006) Post-suburban Europe: planning and politics at the margins of Europe's capital cities. Palgrave Macmillan, Basingstoke
45. Vidal M, Domènech E, Saurí D (2011) Changing geographies of water-related consumption: residential swimming pools in suburban Barcelona. Area 43:67–75
46. Tomao A, Quatrini V, Corona P, Ferrara A, Lafortezza R, Salvati L (2017) Resilient landscapes in Mediterranean urban areas: understanding factors influencing forest trends. Environ Res 156:1–9
47. Francaviglia R, Di Bene C, Farina R, Salvati L, Vicente-Vicente JL (2019) Assessing "4 per 1000" soil organic carbon storage rates under Mediterranean climate: a comprehensive data analysis. Mitig Adapt Strateg Glob Change 24:795–818
48. Chorianopoulos I (2003) North–south local authority and governance differences in the EU networks. Eur Plan Stud 11:671–695
49. Chorianopoulos I, Pagonis T, Koukoulas S, Drymoniti S (2010) Planning, competitiveness and sprawl in the Mediterranean city: the case of Athens. Cities 27:249–259
50. Leontidou L (2012) Athens in the Mediterranean 'movement of the piazzas': spontaneity in material and virtual spaces. City 16:299–312
51. Muñoz F (2003) Lock living: urban sprawl in Mediterranean cities. Cities 20:381–385
52. Vardopoulos I, Tsilika E, Sarantakou E, Zorpas AA, Salvati L, Tsartas P (2021) An integrated SWOT-PESTLE-AHP model assessing sustainability in adaptive reuse projects. Appl Sci 11:7134
53. Glaster G, Hanson R, Ratcliffe MR, Wolman H, Coleman S, Freihage J (2001) Wrestling sprawl to the ground: defining and measuring an elusive concept. Hous Policy Debate 12:681–717
54. Portnov BA, Safriel UN (2004) Combating desertification in the Negev: dryland agriculture vs. dryland urbanization. J Arid Environ 56:659–680
55. Durà-Guimera A (2003) Population deconcentration and social restructuring in Barcelona, a European Mediterranean city. Cities 20:387–394
56. Seifollahi-Aghmiuni S, Kalantari Z, Egidi G, Gaburova L, Salvati L (2022) Urbanisation-driven land degradation and socioeconomic challenges in peri-urban areas: insights from Southern Europe. Ambio 51:1446–1458
57. Arapoglou VP, Sayas J (2009) New facets of urban segregation in Southern Europe. Eur Urban Reg Stud 16:345–362
58. Cassiers T, Kesteloot C (2012) Socio-spatial inequalities and social cohesion in European cities. Urban Stud 49:1909–1924
59. Musterd S, Ostendorf W (1998) Urban segregation and the welfare state: inequality and exclusion in Western cities. Routledge, London
60. Scott AJ (2001) Global city-regions: trends, theory, policy. Oxford University Press, Oxford
61. Maloutas T (2007) Segregation, social polarization and immigration in Athens during the 1990s: theoretical expectations and contextual difference. Int J Urban Reg Res 31:733–758
62. Maloutas T (2007) Socio-economic classification models and contextual difference: the "European socio-economic classes" (ESeC) from a South European angle. South Eur Soc Polit 12:443–460

63. Maloutas T, Karadimitriou N (2001) Vertical social differentiation in Athens: alternative or complement to community segregation. Int J Urban Reg Res 25:699–716
64. King R, Proudfoot L, Smith B (1997) The Mediterranean: environment and society. Arnold, London
65. Kourliouros E (2003) Reflections on the economic–noneconomic debate: a radical geographical perspective from the European South. Antipode 35:781–799
66. Williams AM (1984) Southern Europe transformed: political and economic change in Greece, Italy. Portugal and Spain. Harper & Row, London
67. Delladetsima P (2006) The emerging property development pattern in Greece and its impact on spatial development. Eur Urban Reg Stud 13:245–278
68. Economou D, Petrakos G, Psycharis Y (2007) Urban policy in Greece. In: Van den Berg L, Braun E, Van der Meer J (eds) National policy responses to urban challenges in Europe. Ashgate, Aldershot
69. Souliotis N (2013) Cultural economy, sovereign debt crisis and the importance of local contexts: the case of Athens. Cities 33:61–68
70. Longhi C, Musolesi A (2007) European cities in the process of economic integration: towards structural convergence. Ann Reg Sci 41:333–351
71. Giaccaria P, Minca C (2010) The Mediterranean alternative. Prog Hum Geogr 35:345–365
72. Coccossis H, Economou D, Petrakos G (2005) The ESDP relevance to a distant partner: Greece. Eur Plan Stud 13:253–264
73. Costa F, Noble AG, Pendleton G (1991) Evolving planning systems in Madrid, Rome and Athens. GeoJournal 24:293–303
74. Kresl PK (2007) Planning cities for the future: the successes and failures of urban economic strategies in Europe. Edward Elgar, Cheltenham
75. Kosmas C, Detsis V, Karamesouti M, Kounalaki K, Vassiliou P, Salvati L (2015) Exploring long-term impact of grazing management on land degradation in the socio-ecological system of Asteroussia Mountains, Greece. Land 4:541–559
76. Quaranta G, Salvia R, Salvati L, Di Paola V, Coluzzi R, Imbrenda V, Simoniello T (2020) Long-term impacts of grazing management on land degradation in a rural community of Southern Italy: depopulation matters. Land Degrad Dev 31:2379–2394
77. Sallustio L, Pettenella D, Merlini P, Romano R, Salvati L, Marchetti M, Corona P (2018) Assessing the economic marginality of agricultural lands in Italy to support land use planning. Land Use Policy 76:526–534
78. Vaiou D (1997) Facets of spatial development and planning in Greece. Eur Plan Stud 5:431–433
79. Beriatos E, Gospodini A (2004) Glocalising" urban landscapes: Athens and the 2004 Olympics. Cities 21:187–202
80. Ezcurra R, Rapun M (2006) Regional disparities and national development revisited: the case of Western Europe. Eur Urban Reg Stud 13:355–369
81. Ferrara A, Kelly C, Wilson GA, Nolè A, Mancino G, Bajocco S, Salvati L (2016) Shaping the role of 'fast' and 'slow' drivers of change in forest-shrubland socio-ecological systems. J Environ Manage 169:155–166
82. Garcia-López MA, Muniz I (2010) Employment decentralisation: polycentricity or scatteration? The case of Barcelona. Urban Stud 47:3035–3056
83. Gospodini A (2009) Post-industrial trajectories of Mediterranean European cities: the case of post-Olympics Athens. Urban Stud 46:1157–1186
84. Incerti G, Feoli E, Salvati L, Brunetti A, Giovacchini A (2007) Analysis of bioclimatic time series and their neural network-based classification to characterise drought risk patterns in South Italy. Int J Biometeorol 51:253–263
85. Kazepov Y (2005) Cities of Europe: changing contexts, local arrangements, and the challenge to urban cohesion. Blackwell, Oxford
86. Lambooy JG (1998) Polynucleation and economic development: the Randstad. Eur Plan Stud 6:457–467
87. Lamonica GR, Recchioni MC, Chelli FM, Salvati L (2020) The efficiency of the cross-entropy method when estimating the technical coefficients of input–output tables. Spat Econ Anal 15:62–91

88. Leontidou L, Afouxenidis A, Kourliouros E, Marmaras E (2007) Infrastructure-related urban sprawl: mega-events and hybrid peri-urban landscapes in Southern Europe. In: Couch C, Leontidou L, Petschel-Held G (eds) Urban sprawl in Europe: landscapes, land-use change and policy. Blackwell, Oxford
89. López-i-Villanueva C, Pujadas RI, Carrasco JB (2013) Households within the residential mobility process: the case of the Barcelona metropolitan region. Arch Studi Urb Reg 28:57–84
90. Mancino G, Nolè A, Salvati L, Ferrara A (2016) In-between forest expansion and cropland decline: a revised USLE model for soil erosion risk under land-use change in a Mediterranean region. Ecol Indic 71:544–550
91. Munafò M, Norero C, Sabbi A, Salvati L (2010) Soil sealing in the growing city: a survey in Rome, Italy. Scott Geogr J 126:153–161
92. Salvati L, De Rosa S (2014) 'Hidden polycentrism' or 'subtle dispersion'? Urban growth and long-term sub-centre dynamics in three Mediterranean cities. Land Use Policy 39:233–243
93. Salvati L, Zitti M, Sateriano A (2013) Changes in the city vertical profile as an indicator of sprawl: evidence from a Mediterranean region. Habitat Int 38:119–125
94. Salvati L, Tombolini I, Perini L, Ferrara A (2013) Landscape changes and environmental quality: the evolution of land vulnerability and potential resilience to degradation in Italy. Reg Environ Change 13:1223–1233
95. Salvati L, Zitti M (2009) Substitutability and weighting of ecological and economic indicators: exploring the importance of various components of a synthetic index. Ecol Econ 68:1093–1099
96. Saurí D (2003) Lights and shadows of urban water demand management: the case of the metropolitan region of Barcelona. Eur Plan Stud 11:229–243
97. Torrens PM (2008) A toolkit for measuring sprawl. Appl Spat Anal Policy 1:5–36

Chapter 2
Settlement Expansion and the Formation of Urban Voids

Abstract This contribution examines the concept, significance, and spatial dimension of urban voids, defined as land plots resulting from abandonment, or abrupt change in use. Urban voids acquire meaning through their relationships with surrounding structures, symbolic values, and social functions. The notion of void extends beyond simple emptiness, incorporating relational and topological dimensions—where 'interspaces' become mediums for urban connectivity and scalar transition from architectural to landscape levels. Philosophical perspectives frame voids as both substance and relational space, highlighting their role in shaping human perceptions, dwellings, and spatial organization of cities. With this perspective in mind, voids became places of potential interaction, visual connection, and functional articulation within metropolitan landscapes. Architectural examples, such as squares and streets, illustrate diversified functions. Markets, gardens, and celebratory squares, for instance, serve commerce, social interactions, and symbolic representation, reflecting historical and contemporary urban practices. Pedestrianization of streets further demonstrates the functional adaptation of voids, improving neighborhoods' quality, fostering human-scale experiences, and linking pedestrian systems with public transport. Based on these premises, our work underscores the interplay between form, function, and perception of urban voids, emphasizing their role as connective, dynamic spaces that (i) structure the city, (ii) mediate social interaction, and (iii) integrate multiple scales from micro-architecture to metropolitan (and landscape) levels. By integrating philosophical, architectural, and spatial planning perspectives, urban voids are framed as essential components of sustainable, human-centered metropolitan development.

Keywords Dwelling · Plot · Space · Architecture · Morphology

L. Salvati and I. Konaxis, *Vacant Land*,
SpringerBriefs in Architectural Design and Technology,
https://doi.org/10.1007/978-981-95-4204-8_2

2.1 Introduction

Urban voids can be defined as a (more or less extended) area (namely, land plot or parcel) that becomes available due to obsolescence, abandonment, or a change in its intended (or consolidated) use—either economic or ecological. Such areas may also be referred to abandoned soils, or derelict lands, just to mention relevant examples of soil use [1]. In architectural terms, voids may encompass squares, parks, streets, unbuilt plots, or any open space regardless of scale [2]. These areas are usually recognized for the varying degrees of symbolic, functional, or activity-related value they possess. Urban voids can also create a perceptual void, which arises, for instance, when the distance between two buildings is too large to maintain a direct spatial relationship [3]. These voids may be interpreted as 'internal spaces' within the city, conceptualized as 'concave spaces', enabling the definition of a qualitative field shaped by the boundaries shared with the surrounding objects. A topological approach allows the inclusion of all morphological layers, considering not only their traces but also how they are perceived in contemporary metropolitan landscapes [4]. Topology, in this sense, allows a move beyond fixed forms to grasp the essential qualities inherent in spatial configurations (Fig. 2.1).

Fig. 2.1 Small land plots with no use in-between residential buildings. *Source* authors' photographic archive

The concept of void, in its architectural application, should therefore be understood relationally: it does not exist in isolation but derives meaning from the intervals between built elements. This aspect is captured with terms such as 'interspace'—which emphasize the relation between open (spatial) forms and constructed architectural objects [5]. Unlike simple intervals, the notion of 'interspace' does not refer to a fixed dimensional scale and can extend over variable spatial ranges. It reflects the kaleidoscope of relationships within the urban fabric [6], where potential extends beyond individual architectural units to encompass broader connections with the landscape scale.

From an epistemological perspective, voids can be regarded as a form of 'space-substance'. This is not perfectly distinct from the constituting material bodies—in turn forming a structured field that assumes a positive form relative to solid objects. In contrast, the 'void' as a background assumes a different density, acting neither as 'nothing' nor as 'substance', but as a field mediating forces between objects. Modern physics further complicates this understanding: contemporary post-quantum and post-relativistic frameworks emphasize the inseparability of object and space, highlighting the continuous dynamism and transformation inherent in all things (e.g. [7]). In such a view, distinctions between form and void dissolve, as space functions as an interconnected field within which creation and destruction occur perpetually. The void, therefore, is not an absence, but a generative and relational field underlying the structure and organization of all matter.

2.2 The Spatial Dimension of Urban Voids

From a philosophical perspective, space can be interpreted in two ways. The first considers space *a-priori*, namely a territory where objects are situated, and where their size is pre-established [8]. This interpretation aligns with the Platonic concept of 'void' as a substantive condition, not merely an 'intervallum', but rather the framework that allows bodies to manifest in three dimensions, providing the conditions for their emergence without imposing obstacles. Its Aristotelian evolution views space as a collection of finite places, each defined by dimensional and qualitative parameters (e.g. [9]). The second interpretation extends beyond purely dimensional and geometric characteristics typical of the Cartesian model, incorporating perspectives such as that of the observer, thereby acknowledging relational and perceptual dimensions of space [10].

In this perspective, space embodies two fundamental values, namely compositional action and empty spatiality. The interplay between objects and space is intrinsic, and the act of creating space—clearing, delineating, and structuring it—reveals the human capacity for habitation. 'Space' is intimately associated with the concept of 'place', understood as a domain enabling settlements and dwellings [11]. Rather than a technical or arbitrary void to be filled, space is the organization and harmonization of elements in mutual and contextual relation. Place emerges as the collective arrangement, where objects and humans co-belong. The characteristics of place are inseparable from the inhabiting objects. In this context, voids function as a relational entity, a site of architectural prominence relative to both solids and human movements, interactions, and experiences. Every design presupposes a place, and the place itself is configured in accordance with the intentions and structures of that design [12].

Space, conceptualized through networks of relational interactions, and extending toward broader (open) land parcels, allows for architectural co-presence across multiple scales. Spatial understanding proceeds hierarchically, from simple forms to complex configurations [4]. A city, for instance, can be intended as a layered system of distinguishable neighborhoods, subdivided into streets, squares, and buildings, each with internal partititions forming a spatial organization that extends to the arrangement of individual furnishings (e.g. [13]), which collectively define coherent spatial systems (Fig. 2.2).

From an architectural perspective, examples of urban voids can be identified through their morphological, typological, and functional characteristics, with squares and streets representing paradigmatic instances of emptiness. Especially squares have historically constituted the core of urban life. During the Renaissance, squares were integrated with the urban fabric, forming geometrically defined areas that reflected emerging civic identities (e.g. [14]). In the eighteenth and nineteenth centuries, aesthetic concerns and visual representation often superseded social function, with squares serving as scenic backdrops that emphasized prominent secular or religious buildings [15]. Concurrently, the increasing presence of vehicular traffic necessitated a reorganization of urban movement, which often reduced squares from active social centers to junctions regulating circulation, prefiguring the functional transformations observed in twentieth-century urban design [16].

Contemporary squares rarely retain a single, fixed function. Streets and squares operate as articulations of public space, and their social roles frequently evolve into marketplaces or centers of commercial and leisure activity. Specific typologies have emerged: the market square, primarily oriented toward commerce and movement; the garden square, serving as the internal communal core of residential clusters and integrating greenery, seating, and playing areas; and the celebratory or monumental square, which emphasizes symbolic or collective functions, often centered around significant buildings, monuments, or urban landmarks [17].

In modern implementations, market-oriented squares may integrate pedestrian circulation with retail activity, mirroring the dynamics of shopping centers, while garden squares balance paved and landscaped zones to accommodate diverse uses. Celebratory squares maintain a historical continuity, providing ceremonial and communal significance through spatial organization and design elements such as monuments, fountains, or carefully composed paving patterns. Through these variations, urban voids demonstrate both functional flexibility and the capacity to structure social interactions, reinforcing the concept of open spaces as relational and connective elements within the built-up fabric (e.g. [18]). They exemplify how voids serve not merely as an 'absence', but as a dynamic, inhabitable environment that mediates movement, interaction, and collective experience (Fig. 2.3).

Fig. 2.2 In-between 'physical' essence and 'geometric' regularity: 'monumental' voids within an old cemetery. *Source* authors' photographic archive

2.2.1 Toward an 'Architecture' of Urban Voids?

Being cited extensively in previous discussions, squares are widely considered one of the earliest forms of public space, originating from the organization of residential structures around an open area, in turn representing a sort of void compared with the surrounding 'compact' fabric [19]. This configuration enabled both maximum oversight of internal spaces and defensive advantages against external threats, by minimizing the exposed surface of the surrounding buildings [20]. Courtyards established in this manner often carried symbolic meanings and were adopted as central layouts for various types of worship spaces, including agoras, forums, cloisters, and mosque courtyards. A systematic study of the square's morphology allows for a detailed typology based on variations in angles, side lengths, and the interplay of

Fig. 2.3 Mixed public spaces with squares transformed in parking lots and abandoned green areas acting as 'implicit' urban voids. *Source* authors' photographic archive

formal elements. Further analysis identifies the visual and compositional aspects of squares, including building positioning, facade articulation, openings, porticoes, the integration of greenery, and the interaction between historic and modern architectural forms (e.g. [21]). The relationship between entry points and the square itself is also critical, allowing both historic and contemporary spaces to be interpreted as individual elements or as part of larger compositional systems [22].

Contemporary urban squares generally reflect three primary functional types: commercial, residential, and commemorative (or symbolic). Market squares, evolving from nineteenth-century examples originally designed for visual representation or civic organization, now accommodate commercial and leisure activities [23]. These spaces act as catalysts for movement and social interaction, as observed in prominent European and American urban contexts. The configuration often includes pedestrian-oriented areas that encourage leisure circulation, social interaction, and consumer activity within safe and aesthetically pleasing environments [24]. Garden squares serve as central elements within residential communities, combining recreational, social, and symbolic functions. These spaces vary in scale, degree of enclosure, and the presence of functional elements such as benches, greenery, or play areas, often reflecting local demographic, architectural, and environmental characteristics [25]. The interplay between paved and vegetated areas remains a common motif, producing both functional and aesthetic contrasts (Fig. 2.4).

Fig. 2.4 The interplay between paved and vegetated areas in a residential area downtown. *Source* authors' photographic archive

Commemorative or celebratory squares evoke historical conceptions of urban squares, frequently distinguished from their surrounding built environment. These spaces often act as symbolic centers, highlighting civic identity, communal memory, or monumental structures. They can accommodate public institutions, transport hubs, or extensions of architectural complexes, forming a spatial dialogue between built and open elements. Essential design aspects, including the treatment of paving, the positioning of monuments, and visual counterpoints, reinforce the integrative function of these public spaces [26]. Squares located in front of civic (or religious) buildings exemplify the relationship between communal activity and architectural symbolism, often structured to accommodate multiple functions while maintaining spatial coherence and visual prominence (Fig. 2.5).

2.3 Urban Voids, Compact Cities, and Public Space

The 'pedestrianization' of urban streets became a significant topic of debate through the mid-twentieth century, particularly following the initial closures of inner spaces to vehicular traffic. These interventions were initially motivated with two main objectives: addressing the growing congestion in central urban areas, and safeguarding

Fig. 2.5 Historical buildings, monumental green spaces, and blue infrastructures. *Source* authors' photographic archive

the most emblematic monumental spaces from vehicular intrusion, while promoting pedestrian use of historic city centers, whose conservation and revitalization were central concerns in contemporary planning discourse. Over time, pedestrianization evolved with varying approaches and methodologies, depending on specific urban contexts and spatial scales [27]. This transcends a mere regulatory practice to become a recognized dimension of 'urban design' culture, reflecting a broader concern for the quality and usability of public space [28].

Pedestrian areas function as dedicated public spaces, encompassing streets, squares, open spaces, and alleys exclusively reserved for pedestrian use. These areas may result either from the conversion of previously vehicular streets or from routes originally designed for pedestrian circulation. Unlike streets temporarily closed to vehicles, pedestrian zones entail a complete spatial reallocation [29]. Pedestrian areas can be classified according to local contexts: (i) those within established urban fabrics, whether historic or architecturally modest yet representative,(ii) directional areas dominated by tertiary activities in central and peripheral zones; and (iii) marginal areas serving recreational, landscape, or residential functions [30]. They may also be differentiated by the spatial organization of pedestrian and vehicular flows: at-grade separation, where pedestrians, traffic, and parking coexist on different surfaces, or vertical separation, with vehicles and parking below and pedestrian paths above, often connected by elevated plates linking urban elements—a model exemplified by early pedestrian zones such as New York's Rockefeller Center. Variants include sunken squares, with roadways on the upper level and pedestrian spaces below, as seen e.g. in the Forum des Halles in Paris.

Today, the topic retains its relevance, albeit with less controversy and often distinct objectives. These targets include enhancing urban mobility by prioritizing public transport over private vehicles, particularly in European contexts, and improving the overall environmental and social quality of metropolitan areas. In some instances, pedestrianization has facilitated the reevaluation of building replacement policies, restoring the integrity of historical urban fabrics that had been disrupted by prior development strategies [31]. In American cities, where interventions such as highway construction and large-scale commercial developments often failed to generate attractive public environments (e.g. [32]), pedestrianization has been instrumental in reclaiming urban spaces for human use. For instance, pedestrian initiatives in Boston have restored the historical character of 19th- and early 20th-century streets, creating accessible, human-scaled environments. Similarly, the Washington Street Mall reinstated the street's original appearance, distinguishing pedestrian areas from vehicular routes and expanding sidewalk space [33]. Street furnishings—including both historically styled and contemporary lighting, benches, telephone booths, and bollards—are carefully designed to align with the intended urban scene. Transparent shelters have been installed in certain locations to demarcate pedestrian zones while maintaining visual continuity and balancing the vertical presence of adjacent buildings.

The implementation of pedestrianized axes requires careful analysis of existing traffic flows to identify feasible alternatives. Within this framework, public transportation is essential for maintaining accessibility to a large number of users. The development of central pedestrian zones in major cities has frequently depended on the availability of rapid transit systems, particularly subways, whose routes often correspond with pedestrianized areas. In certain cases, subway operators have contributed significantly to the cost of pedestrianization projects.

2.3.1 Vacant Land as a Relational Linkage Between Architecture and Nature

From a panoramic perspective, cities appear as coherent, static elements within the landscape, such as a hilltop village or a historic center viewed from afar [34]. However, once navigated through its road network, the urban environment fragments into multiple visual sequences. Roads, as conduits of movement and observation, allow the perception of buildings, architectural features, furnishings, and overall urban structure from continuously shifting vantage points—producing a dynamic experience. In contrast, squares—through their spatial expansion—encourage pause and reflection, offering comprehensive visual perspectives from various observation points.

With this perspective in mind, the intimate and complex morphology of voids—understood not as open space in isolation but as the spatial figure generated through the morphological relationship between solids—requires an effective and sophisticated framework analyzing the 'connective' mechanisms between architectural elements [35]. Architecture, in this perspective, is not considered solely as an autonomous object, but rather in relation to the pre-existing urban fabric and the broader territorial context (e.g. [36]). A deep reflection on the configuration of open space as forms of the void, constructed through relational interactions with architectural volumes, contributes to the understanding of potential figures for open spaces.

In many contemporary contexts, urban space is conceived primarily as a conduit for functional circulation, frequently neglecting the complexity of social interactions and leading to the erosion of localized forms of urban life. The continuous expansion of metropolitan areas exacerbates this challenge, resulting in spaces devoid of identity, historical memory, or social connectivity. Urban voids often become marginal zones, e.g. as 'museified' public spaces that acquire significance only through their relation with the surrounding urban fabric—yet remaining socially sterile. In contemporary suburban environments, genuine public space is increasingly absent, often replaced by commercial hubs, while individual identification through social interaction diminishes [37]. To counteract this, voids should be envisioned as multifunctional public spaces that support personal and collective life. They should enable activities of leisure, interaction, and social aggregation, potentially enhanced through the integration of landscape elements such as water features and greenery. Playful architectural interventions can further enrich these spaces, fostering engagement, creativity, and novel (relational) dynamics within complex, contemporary urban environments.

2.4 (Re)connecting Urban Design with the Ecological Functions of Vacant Land

Moving to broader spatial scales, cities can be understood as a population cluster occupying a finite territory, consisting of buildings and the spaces between them, forming a complex system of formal and functional structures coordinated according with social dynamics [38]. Cities function as centers of coordination for wider regions, organizing population, buildings, and open spaces into coherent geographical units [39]. By contrast, rural landscapes are characterized by human-shaped agricultural or semi-natural spaces, distinct from unmodified natural environments such as deserts or glaciers [40]. The agricultural landscape represents an ongoing interaction between human activity and natural terrain, being increasingly transformed by urban expansion (Fig. 2.6).

Historically, urban and rural spaces have maintained distinct social, economic, and spatial organizations [41]. Urbanization initially referred to the concentration of populations around central nodes, and has rapidly evolved to include the diffusion of

Fig. 2.6 Contested fringe land under conversion from agriculture to residences. *Source* authors' photographic archive

'urban' cultural practices and social norms into the countryside [42–44]. While urban culture extends beyond city limits, the distinction between urban and rural settings remains [45–47]. Contemporary rural societies have diversified substantially under industrialization, with cities exerting a growing influence over agricultural regions through service provision, market functions, and socioeconomic integration [48–50].

Historically, cities were often distinguished from the countryside by walls enclosing the inhabited areas, while linear suburbs extended along access roads outside the gates [51]. The countryside, in contrast, was primarily agricultural, with clearly demarcated landscapes and human activities [52]. In many Western—particularly Mediterranean—cultures, the city has been conceptualized in opposition to the countryside, representing a contained, protected, and enclosed space intended to serve collective needs, both historically and within the collective imagination [53]. Modern urban expansion, driven by rising land values, has reinforced this division while simultaneously encouraging the proliferation of vacation homes and secondary residences [54]. The creation of parks and gardens has served as a mechanism to reconcile urban and rural environments, providing functional, recreational, and aesthetic integration, mitigating urban heat through evapotranspiration, enhancing opportunities for physical activity, and offering areas for passive contemplation and natural experience [55–57]. The coexistence of active recreational zones with naturalistic areas reflects the value of layered and multifunctional green spaces within cities.

2.4.1 The Role of Urban Greening

Mixing urban and rural settings, metropolitan environments are composed of both natural and anthropogenic elements, the latter resulting from human activity, all of which are subject to continuous transformation due to natural phenomena such as earthquakes and floods, as well as through deliberate human interventions that continuously alter the appearance of the earth's surface. The modification of the earth's surface is driven by endogenous forces, which originate within the earth's crust, as exemplified by volcanic eruptions and seismic events, and by exogenous forces acting externally. On the one hand, natural elements, which constitute non-humanized environments, include air, atmospheric conditions, water in its various forms such as rivers, lakes, streams, and seas, as well as soils and the subsoil, characterized by diverse components including rock, gravel, sand, and minerals [58–60].

Anthropogenic elements, on the other hand, progressively transform the original character of landscapes over time [61]. These include cultivated landscapes, where spontaneous vegetation has been converted into systematically organized production, deforestation works that alter the distribution of plant species, and hydraulic interventions such as embankments, irrigation, and reclamation projects, which modify water regimes and reshape natural habitats [62]. Human constructions—from residential buildings and religious structures to agricultural, industrial, and infrastructural developments including roads, bridges, railways, airports, ports, and large-scale utilities—further contribute to the transformation of the landscape, gradually becoming integrated and defining a mostly 'humanized' environment [63].

Within the urban context, greenery performs multiple interrelated functions. Green areas mitigate the impact of urbanization by regulating microclimates, reducing summer heat through evapotranspiration, and counteracting environmental degradation. They also contributes to health and well-being, particularly near hospitals, where the presence of well-maintained greenery and aromatic plant species can aid patient convalescence both physically and psychologically (Fig. 2.7). Vegetation also provides protective functions by stabilizing vulnerable areas, such as riverbanks and slopes, while its removal can exacerbate environmental fragility. Green spaces also foster recreation, leisure, and community interaction, allowing urban areas to maintain human-scale livability. These areas also serve cultural and educational purposes, facilitating the dissemination of botanical and environmental knowledge, preserving historical gardens and significant plant specimens as living cultural monuments, and promoting engagement with nature across generations [31].

From an aesthetic and architectural perspective, the integration of greenery enhances metropolitan landscapes, rendering cities more pleasant and improving the relationship between built structures and their surroundings. Water, as an essential

Fig. 2.7 Vacant land, greening, and a suburban hospital. *Source* authors' photographic archive

condition for human life, has historically shaped the formation of settlements [64]. Early human communities predominantly arose in temperate and subtropical river basins, where irrigation and riverine management were fundamental to sustaining complex civilizations. The distribution of water resources highlights the intrinsic connection between human societies and their natural environment.

Contemporary urban greenery has finally emerged as a critical venue for citizens' leisure, recreation, sports, and cultural activities, diminishing the historical distinction between formal, 'structured' parks and informal, 'naturalistic' gardens. These spaces aim to counteract congestion derived from settlement compactness, reduce the expansion of secondary residences, and mitigate air pollution. However, social dynamics and changing patterns of land use have sometimes limited the effectiveness of parks, with challenges including vandalism, underuse, crime, and social exclusion. Conversely, many urban green areas successfully support social interaction, relaxation, and recreation, contributing to the vitality of the city and its human-scale livability [65].

2.5 Concluding Remarks

By debating on the continuous interplay between architectural forms and natural functions, we elucidated different modalities through which architecture engages in a dialectical relationship with nature. This interaction can be understood both in terms of architecture situated within the natural context, and in terms of the integration of natural elements within architectural design. In the first case, architecture assumes a proactive role: the focus is placed on how architectural interventions can constitute a qualified landscape and how they can harmoniously integrate into existing natural environments. In the second perspective, architecture assumes a more receptive position, becoming the lens through which one examines the inclusion of natural elements within architectural projects, whether applied to façades, interiors, or other structural components.

Throughout these cases, architecture remains the primary subject of study, yet its interaction with natural elements is considered in a manner that emphasizes analysis and observation rather than intervention. However, in a final 'thematic' perspective, the attention returns to architecture as a central object, being assumed as an autonomous and non-contextualized entity, with the focus placed on the study of urban morphology, rather than its interaction with surrounding natural features. This framework provides a comprehensive approach to understanding the multifaceted ways in which architecture and nature coexist, intersect, and mutually co-evolve.

References

1. Alves D, Barreira AP, Guimarães ME, Panagopoulos T (2016) Historical trajectories of currently shrinking Portuguese cities: a typology of urban shrinkage. Cities 52:20–29
2. Souliotis N (2013) Cultural economy, sovereign debt crisis and the importance of local contexts: the case of Athens. Cities 33:61–68
3. Daly G, Kitchin R (2013) Shrink smarter? planning for spatial selectivity in population growth in Ireland. Administration 60(3):159–186
4. Salvati L, Morelli VG, Rontos K, Sabbi A (2013) Latent exurban development: City expansion along the rural-to-urban gradient in growing and declining regions of southern Europe. Urban Geogr 34(3):376–394
5. Fonseca ML (2008) New waves of immigration to small towns and rural areas in Portugal. Popul Space Place 14(6):525–536
6. Champion T, Hugo G (2003) New forms of urbanization: Beyond the urban-rural dichotomy. Routledge, London
7. Arapoglou VP, Sayas J (2009) New facets of urban segregation in southern Europe—gender, migration and social class change in Athens. European Urban and Regional Studies 16(4):345–362
8. Storper M (2011) Why do regions develop and change? The challenge for geography and economics. J Econ Geogr 11:333–346
9. Amato F, Maimone BA, Martellozzo F, Nolè G, Murgante B (2016) The effects of urban policies on the development of Urban areas. Sustainability 8:297
10. Chorianopoulos I, Pagonis T, Koukoulas S, Drymoniti S (2010) Planning, competitiveness and sprawl in the Mediterranean city: The case of Athens. Cities 27:249–259

11. Bhatta B (2010) Causes and consequences of urban growth and sprawl. Analysis of urban growth and sprawl from remote sensing data. Springer, Berlin
12. Maloutas T (2004) Segregation and residential mobility: spatially entrapped social mobility and its impact on segregation in Athens. Eur Urban RegNal Stud 11(3):195–211
13. Sato Y, Yamamoto K (2005) Population concentration, urbanization, and demographic transition. J Urban Econ 58(1):45–61
14. De Rosa S, Salvati L (2016) Beyond a 'side street story'? Naples from spontaneous centrality to entropic polycentricism, towards a 'crisis city.' Cities 51:74–83
15. Mallach A, Haase A, Hattori K (2017) The shrinking city in comparative perspective: contrasting dynamics and responses to urban shrinkage. Cities 69:102–108
16. Matanle P, Sato Y (2010) Coming soon to a city near you! Learning to live 'beyond growth' in Japan's shrinking regions. Soc Sci Jpn J 13(2):187–210
17. Montgomery MR (2008) The urban transformation of the developing world. Science 319(5864):761–764
18. Pili S, Grigoriadis E, Carlucci M, Clemente M, Salvati L (2017) Towards sustainable growth? a multi-criteria assessment of (changing) urban forms. Ecol Ind 76:71–80
19. Panori A, Psycharis Y, Ballas D (2019) Spatial segregation and migration in the city of Athens: investigating the evolution of urban socio-spatial immigrant structures. Population, Space and Place 2209
20. Porter JR, Howell FM (2009) On the 'Urbanness' of metropolitan areas: testing the homogeneity assumption, 1970–2000. Popul Res Policy Rev 28:589–601
21. Salvati L, Ranalli F, Carlucci M, Ippolito A, Ferrara A, Corona P (2017) Forest and the city: a multivariate analysis of peri-urban forest land cover patterns in 283 European metropolitan areas. Ecol Ind 73:369–377
22. Giannakourou G (2005) Transforming spatial planning policy in Mediterranean countries: Europeanization and domestic change. Eur Plan Stud 13:319–331
23. Gospodini A (2006) Portraying, classifying and understanding the emerging landscapes in the post-industrial city. Cities 23(5):311–330
24. Johnson DL, Lewis LA (2007) Land degradation—creation and destruction. Rowman & Littlefield, Lanham
25. Gospodini A (2009) Post-industrial trajectories of Mediterranean European cities: the case of post-Olympics Athens. Urban Studies 46:1157–1186
26. Rink D, Haase A, Grossmann K, Couch C, Cocks M (2012) From long-term shrinkage to re-growth? the urban development trajectories of Liverpool and Leipzig. Built Environment 38(2):162–178
27. Wiechmann T, Pallagst KM (2012) Urban shrinkage in Germany and the USA: a comparison of transformation patterns and local strategies. Int J Urban Reg Res 36(2):261–280
28. Todes A, Kok P, Wentzel M, Van Zyl J, Cross C (2010) Contemporary South African urbanization dynamics. Urban Forum 21(3):331–348
29. Beriatos E, Gospodini A (2004) Glocalising urban landscapes: Athens and the 2004 olympics. Cities 21:187–202
30. Weaver R, Bagchi-Sen S, Knight J, Frazier AE (2016) Shrinking cities: understanding urban decline in the United States. Routledge, London
31. Haase A (2011) Residential change and demographic challenge: the inner city of East Central Europe in the 21st century. Ashgate, Aldershot
32. Chorianopoulos I, Tsilimigkas G, Koukoulas S, Balatsos T (2014) The shift to competitiveness and a new phase of sprawl in the Mediterranean city: enterprises guiding growth in Messoghia—Athens. Cities 39:133–143
33. Van Deusen JR (2002) Public space design as class warfare: Urban design, the 'right to the city' and the production of Clinton Square, Syracuse NY. GeoJournal 58(2):149–158
34. Cakir G, Un C, Baskent EZ, Kose S, Sivrikaya F, Keles S (2008) Evaluating urbanization, fragmentation and land use/cover change pattern in Istanbul city, Turkey from 1971 to 2002. Land Degrad Dev 19:663–675

35. Amato F, Tonini M, Murgante B, Kanevski M (2018) Fuzzy definition of Rural Urban Interface: An application based on land use change scenarios in Portugal. Environ Model Softw 104:171–187
36. Carlucci M, Grigoriadis E, Rontos K, Salvati L (2017) Revisiting an hegemonic concept: long-term 'Mediterranean Urbanization' in between city re-polarization and metropolitan decline. Appl Spat Anal Policy 10(3):347–362
37. Salvati L, Ferrara C (2013) Do changes in vegetation quality precede urban sprawl? Area 45(3):365–375
38. Bettencourt LM, Lobo J, Helbing D, Kühnert C, West GB (2007) Growth, innovation, scaling, and the pace of life in cities. Proc Natl Acad Sci 104(17):7301–7306
39. Amcoff J, Westholm E (2007) Understanding rural change—demography as a key to the future. Futures 39(4):363–379
40. Serra P, Vera A, Tulla AF, Salvati L (2014) Beyond urban-rural dichotomy: exploring socioeconomic and land-use processes of change in Spain (1991–2011). Appl Geogr 55:71–81
41. Seto KC, Ramankutty N (2016) Hidden linkages between urbanization and food systems. Science 352(6288):943–945
42. Haase A, Rink D, Grossmann K, Bernt M, Mykhnenko V (2014) Conceptualizing urban shrinkage. Environ Plan A 46(7):1519–1534
43. Salvati L, Tombolini I, Gemmiti R, Carlucci M, Bajocco S, Perini L, Ferrara A, Colantoni A (2015) Complexity in action: untangling latent relationships between land quality, economic structures and socio-spatial patterns in Italy. PLoS ONE 12(6):e0177853
44. Tsilimigkas G, Kizos T (2014) Space, pressures and the management of the Greek landscape. Geografiska Annaler: Series B, Human Geography 96(2):159–175
45. Radeloff VC, Hammer RB, Stewart SI, Fried JS, Holcomb SS, Mckeefry JF (2005) The wildland-urban interface in the United States. Ecol Appl 15(3):799–805
46. Recanatesi F, Clemente M, Grigoriadis S, Ranalli F, Zitti M, Salvati L (2016) A fifty-years sustainability assessment of Italian Agro-forest Districts. Sustainability 8(1):32
47. Salvati L, Carlucci M (2011) The economic and environmental performances of rural districts in Italy: are competitiveness and sustainability compatible targets? Ecol Econ 70(12):2446–2453
48. Portnov BA, Safriel UN (2004) Combating desertification in the Negev: dryland agriculture vs. dryland urbanization. J Arid Environ 56:659–680
49. Savo V, Caneva G, McClatchey W, Reedy D, Salvati L (2014) Combining environmental factors and agriculturalists' observations of environmental changes in the traditional terrace system of the Amalfi Coast (Southern Italy). Ambio 43(3):297–310
50. Schewenius M, McPhearson T, Elmqvist T (2014) Opportunities for increasing resilience and sustainability of urban social-ecological systems: insights from the URBES and the cities and biodiversity outlook projects. Ambio 43(4):434–444
51. Champion T (2001) The continuing urban-rural population movement in Britain: trends, patterns, significance. Espace Populations Societes 19(1):37–51
52. Coluzzi R, Bianchini L, Egidi G, Cudlin P, Imbrenda V, Salvati L, Lanfredi M (2022) Density matters? settlement expansion and land degradation in peri-urban and rural districts of Italy. Environ Impact Assess Rev 92:106703
53. Di Feliciantonio C, Salvati L (2015) Southern' alternatives of urban diffusion: investigating settlement characteristics and socioeconomic patterns in three Mediterranean regions. Tijdschr Econ Soc Geogr 106(4):453–470
54. Cuadrado-Ciuraneta S, Durà-Guimerà A, Salvati L (2017) Not only tourism: unravelling suburbanization, second-home expansion and 'rural' sprawl in Catalonia Spain. Urban Geogr 38(1):66–89
55. Colantoni A, Grigoriadis E, Sateriano A, Venanzoni G, Salvati L (2016) Cities as selective land predators? a lesson on urban growth, (un)effective planning and sprawl containment. Sci Total Environ 545–546:329–339
56. Delladetsima PM (2006) The emerging property development pattern in Greece and its impact on spatial development. Eur Urban RegNal Stud 13(3):245–278

57. Reckien D, Martinez-Fernandez C (2011) Why do cities shrink? Eur Plan Stud 19(8):1375–1397
58. Gargiulo Morelli V, Rontos K, Salvati L (2014) Between suburbanisation and re-urbanisation? revisiting the Urban life cycle in a Mediterranean compact city. Urban Research and Practice 7(1):74–88
59. Grekousis G, Manetos P, Photis YN (2013) Modeling urban evolution using neural networks, fuzzy logic and GIS: the case of the Athens metropolitan area. Cities 30:193–203
60. Serra P, Pons X, Saurì D (2008) Land-cover and land-use change in a Mediterranean landscape: a spatial analysis of driving forces integrating biophysical and human factors. Appl Geogr 28(3):189–209
61. Schneider A, Woodcock CE (2008) Compact, dispersed, fragmented, extensive? a comparison of urban growth in twenty-five global cities using remotely sensed data, pattern metrics and census information. Urban Studies 45(3):659–692
62. Smiraglia D, Ceccarelli T, Bajocco S, Salvati L, Perini L (2016) Linking trajectories of land change, land degradation processes and ecosystem services. Environ Res 147:590–600
63. Salvati L (2014) Agro-forest landscape and the 'fringe'city: A multivariate assessment of land-use changes in a sprawling region and implications for planning. Sci Total Environ 490:715–723
64. Gkartzios M (2013) 'Leaving Athens': narratives of counterurbanisation in times of crisis. J Rural Stud 32:158–167
65. Paul V, Haslam Mckenzie F (2013) Peri-urban farmland conservation and development of alternative food networks: insights from a case study area in metropolitan Barcelona (Catalonia, Spain). Land Use Policy 30:94–105
66. Salvati L, Zitti M, Sateriano A (2013) Changes in city vertical profile as an indicator of sprawl: evidence from a Mediterranean urban region. Habitat Int 38:119–125
67. Wiechmann T (2008) Errors expected—aligning urban strategy with demographic uncertainty in shrinking cities. Int Plan Stud 13(4):431–446

Chapter 3
Vacant Land and Urbanization Processes

Abstract By investigating the intimate dynamics of urban–rural interactions, the present contribution envisions the emergence of fragmented (and, possibly, 'void') landscapes within contemporary metropolitan regions, focusing on low-density settlement expansion, green infrastructures, and processes of socioeconomic integration. Urban sprawl has produced heterogeneous and often discontinuous spaces, including abandoned land plots, infrastructural corridors, and other forms of mixed landscapes with negligible value—either economic and ecological. These phenomena reflect the interplay of speculative development, regulatory insufficiencies, and socioeconomic transformations, challenging traditional notions of planning, landscape identity, and environmental stewardship. The contribution also examines patterns of decentralization and socio-spatial segregation, emphasizing the role of peri-urban fringes in ecological balance and economic performances. Particular attention is given to the inherent transformation of urban voids into non-places, which crystallize the tensions between growth-oriented development, cultural heritage conservation, and sustainable land management. The study proposes that vacant and peripheral spaces should be understood as critical lenses for urban analysis, offering insights into the contradictions of modern urbanization and potential avenues for intervention. By integrating historical perspectives with contemporary urban realities, this work definitely underscores the importance of holistic, landscape-sensitive planning approaches that balance ecological integrity, social cohesion, and economic vitality. Ultimately, the findings advocate for a strategic rethinking of metropolitan growth, where the fringe operates not as a residual zone, but as a mediator between city and countryside, capable of fostering resilient, multifunctional, and sustainable urban futures.

Keywords Buildings · Design · Density · Landscape · Public space

L. Salvati and I. Konaxis, *Vacant Land*,
SpringerBriefs in Architectural Design and Technology,
https://doi.org/10.1007/978-981-95-4204-8_3

3.1 Introduction

The polarization between urban and rural territories represents one of the most enduring and influential issues in Western history [1]. This dualism has long shaped the ways in which landscapes are conceptualized and transformed—not merely as contrasting physical environments, but also as opposing symbolic domains [2]. Historically, cities have been associated with density, modernity, innovation, and economic dynamism, whereas countryside has been often linked to continuity, tradition, ecological balance, and cultural heritage [3]. The dialogue—and, occasionally, the antagonism—between these two spheres has been extensively examined, highlighting how urban expansion into rural areas reshapes both material landscapes and cultural imaginaries [4].

Over the past century, intense and continuous urbanization processes have intensified this interaction, producing a mosaic of fragmented and heterogeneous spaces at the periphery of large cities [5–7]. Metropolitan fringes—once a zone of gradual transition between built environments and open countryside—increasingly manifest discontinuities, where agricultural landscapes are intersected by infrastructural corridors, industrial clusters, generic housing developments, and underutilized or abandoned parcels [8]. Such transformations rarely arise from coherent planning logics, instead, they often reflect unintended consequences of speculative development and reactive governance [9]. Consequently, spaces emerge that are poorly integrated into broader territorial strategies and disconnected from socio-cultural and environmental contexts [10].

This dynamic is exemplified by urban sprawl—a low-density, automobile-dependent expansion of residential settlements that consumes large tracts of land with little consideration for ecological limits or community needs [11]. The unchecked proliferation of suburban settlements generates multiple consequences, including the loss of agricultural soils, ecosystem fragmentation, degradation of landscape identity, and the entrenchment of unsustainable mobility patterns [12]. In parallel, the notion of 'non-places'—spaces of temporary occupation devoid of historical depth or cultural anchorage—becomes increasingly relevant, as infrastructural nodes, shopping centers, parking lots, and uniform residential estates transform land into a commodity stripped of memory and meaning [13].

Peri-urban fringes have become privileged sites for speculative practices (Fig. 3.1). Rather than reinvesting in the rehabilitation or adaptive reuse of existing structures, real estate markets and local authorities often prioritize new construction, even amidst demographic stagnation or decline [14]. This trend undermines the conservation of cultural and architectural heritage while accelerating vacancy and obsolescence, reinforcing perceptions of marginality and redundancy [15]. The cumulative effect of such processes is the erosion of both ecological and symbolic values at the rural–urban interface [16]. Landscapes that once embodied productive, cultural, and environmental functions are increasingly reduced to residual, vacant lots, awaiting further cycles of speculation or neglect [17].

Fig. 3.1 Fringe land waiting for edification. *Source* authors' photographic archive

In these contexts, vacant land should not merely be interpreted as a proxy of economic decline, but approached as a critical lens through which contemporary urbanization can be interrogated [18]. It crystallizes the contradictions inherent in growth-oriented planning and exposes the limitations of regulatory systems that fail to balance development with conservation [19]. Thus, vacant land provides a pivotal entry point for analyzing the tension between urban expansion and ecological stewardship, economic imperatives and cultural memory, short-term speculation and long-term sustainability [20].

3.2 Urban–Rural Polarization and Metropolitan Growth

Urban–rural polarizations have profoundly shaped Western culture, and the interactions between city and countryside have long been subjects of rigorous study [21]. The expansion of urban fabrics into rural areas has generated fragmented spaces often lacking coherent planning or design logic [22]. Phenomena such as sprawl, uncontrolled land consumption, economic polarization, and the creation, consolidation, and degradation of non-places, have reshaped peri-urban fringes [23], diminishing environmental values, fragmenting rural landscapes, and privileging

Fig. 3.2 Construction sites and fragmentation of relict fringe landscapes. *Source* authors' photographic archive

speculative development where existing building heritage is inadequately preserved (Fig. 3.2).

There is an increasing urgency to recombine and strengthen these fragile, weakly interconnected territories through integrated approaches that complement meropolitan systems while respecting their economic, social, and demographic role [24]. Maintaining a reciprocal relationship with open spaces enables strategies to counter urban marginalization and social degradation, for instance by fostering networks of agricultural excellence, green infrastructures, and opportunities in hiking, agri-tourism, organic agriculture, and social inclusion [25–27]. In this way, city boundaries transform from a simple line of demarcation into a mediator between urban and rural domains (e.g. [28]). Moving from a divisive, conflictual conception of urban borders toward a more inclusive, mediating notion forms the basis for the present study, addressing low-density sprawl and the mechanisms behind the creation of extensive land voids at the edges of urban fabrics [29].

This contribution additionally employs a diachronic, narrative, and illustrative approach to sequential stages of metropolitan development, seeking to understand the spatial (and logical) connections between consolidated urban cores and widespread settlement diffusion. The focus is on landscapes that are no longer urban yet not fully rural—characterized by unique physiognomies, specific uses, and particular functions. Individual peculiarities are examined in relation to broader territorial patterns, often lacking causal links or sufficient economic, environmental, and social connectivity due to unregulated growth [30]. The ultimate aim is to develop an integrated methodological framework for interventions in suburban void landscapes, supported by case studies where landscape-scale redevelopment is prioritized [31]. We emphasize the central functions of consolidated cities while interpreting city-countryside interdependence across demographic, political, socioeconomic, and cultural dimensions [32].

The 2008 global crisis accentuated urban–rural conflicts, which can only be addressed through sustainable urban planning (e.g. [33]). Contemporary tools for territorial governance facilitate the design of strategic scenarios, promoting interaction, connectivity, and valorization of metropolitan components, including vacant spaces [34]. Empty spaces and their interactions with built and natural environments in suburban areas were characterized as outcomes of temporally (and spatially) distinct processes of settlement expansion and economic growth [35]. By adopting landscape as a structural element of metropolitan fringes, the study explores interdisciplinary contributions and proposes a comparative framework—from experimental diagnosis to design methodology—focused on the intrinsic relationships among economic agents. The goal is to respond to the complexity of territories where nature, economy, society, culture, and governance are inseparably intertwined [36].

In this perspective, urban growth is among the most debated issues in regional sustainability, resulting from the interaction of historical, political, and cultural factors and contributing to a broader discourse on differential policy and planning implications for e.g. European and North American contexts, in light of global socioeconomic dynamics [37]. Earlier studies highlight how urban growth manifests differently across metropolitan areas, requiring detailed comparative investigation of environmental, social, and economic impacts [38]. Historical and diachronic analyses are essential for understanding the 'territorial' complexity inherent in any process of settlement expansion (e.g. [39]). While environmental and landscape contexts are structural to metropolitan transformations (Fig. 3.3), marginal territories—agricultural and forest areas within peri-urban landscapes—play decisive roles in sustaining ecological quality and generating social capital, thereby enhancing territorial competitiveness [40].

3.2.1 The Long-Term Evolution of Metropolitan Regions

Over the past decades, urban areas worldwide have experienced significant population growth alongside rapid, often unplanned, spatial expansion [41–43]. This process has contributed to the development of metropolitan regions in which the traditional distinctions between city centers, suburban areas, and rural landscapes have increasingly blurred [44]. Cities are progressively merging with adjacent urban entities, giving rise to extensive conurbations [45]. With over twenty million inhabitants, the New York metropolitan area exemplifies one of the largest urban agglomerations globally (e.g. [46]). Its spatial extent encompasses a vast array of infrastructures,

Fig. 3.3 Fringe greenways and blueways giving ecological and aestethic value to settlement-farmland interfaces. *Source* authors' photographic archive

industrial zones, airports, parks, commercial centers, and peripheral settlements that were previously morphologically distinct. This spatial expansion has been accompanied by profound socioeconomic transformations: commercial and industrial activities have frequently relocated from central districts to peripheral zones, while central areas have increasingly become residential spaces for higher-income populations, offering access to premium services and cultural amenities [47].

Historically, urban decentralization is not a novel phenomenon. Cities have exhibited dispersed settlement patterns for centuries, often in response to prevailing economic and social conditions [5]. For instance, in the classical era, affluent populations frequently established residences outside core urban areas, resulting in low-density suburban developments [48]. Comparable patterns were observed in non-Western contexts, such as China during the Ming dynasty, highlighting the global and historical dimensions of urban diffusion. These patterns suggest a strong correlation between economic prosperity and the spatial distribution of urban populations: as cities mature economically, they tend to expand outward, with lower population densities in peripheral zones [49–51].

In Europe, the early phases of urban decentralization can be traced to the growth of major cities such as London or Paris. Rapid economic development attracted rural migrants, stimulating outward residential expansion [52]. Well-developed public transport systems facilitated commuting from peripheral areas, enabling wealthier households to reside in low-density settlements while maintaining close social and economic ties to inner cities. Beyond the suburban fringe, exurban areas emerged, being characterized by low-density residential settlements that remained socially and economically integrated with the city despite predominantly rural land uses. As industrialization progressed, similar patterns of urban dispersion emerged across other European cities.

By the late nineteenth and early twentieth centuries, urban sprawl had become a widely recognized phenomenon, particularly in North America [53]. Rapid population growth, combined with affordable public transport, allowed households to settle far from central employment areas. Over time, economic decentralization accompanied residential dispersion, with industries and commercial activities relocating to suburban zones, giving rise to heterogeneous suburban landscapes differentiated by socioeconomic status and urban function [54]. In the mid- to late twentieth century, suburbanization evolved further. Residential developments became larger and more diversified, incorporating self-governing communities, private amenities and, in some cases, gated communities [55]. Suburban models have increasingly combined elements of centralization and decentralization, fostering intensive social and economic interactions while maintaining spatial dispersion [56]. These transformations raised concerns about the sustainability of urban growth, particularly regarding land consumption, environmental impact, and infrastructure demands (Fig. 3.4).

Recent decades have witnessed the global proliferation of large-scale commercial and industrial zones, including shopping centers and service-oriented settlements [57]. These developments often replicate the dualism observed in historical urban cores: centralizing multiple functions to generate economic and social synergies while simultaneously dispersing spatially to optimize accessibility and land-use efficiency [58]. The globalization of such settlement patterns underscores the increasingly interconnected nature of urban development processes, with similar dynamics emerging across Europe, North America, and Asia [59], and socio-environmental consequences spreading in both advanced economies and emerging countries [60].

Population and employment decline in city centers has simultaneously created opportunities for new forms of socioeconomic development (Fig. 3.5). As industries and residents relocate outward, suburban areas increasingly acquire mixed-use characteristics, while central urban districts undergo partial revitalization [61]. This process, often associated with gentrification, involves the return of higher-income households and the introduction of new commercial activities to previously depopulated urban cores [62]. Large-scale urban interventions, such as the redevelopment of central districts for international events, have further reinforced such trends. For instance, the 1992 Barcelona Olympics successfully revitalized depopulated neighborhoods and encouraged the repopulation of central areas, whereas the 2004 Athens Olympics achieved more limited success in terms of restoring historic cores and promoting demographic repolarization [63–65].

In the United States, significant central-area redevelopment began in the late 1960s, exemplified by large-scale projects in downtown Los Angeles. Despite these efforts, suburbanization has continued to dominate residential patterns [66–70].

Fig. 3.4 Land take, infrastructural development and unsustainable urban expansion. *Source* authors' photographic archive

Single-family dwellings have drastically increased in size, growing from an average of approximately 90 m^2 post-World War II to around 230 m^2 by the end of the century. This reflects changing lifestyles and the appeal of suburban living [71–74].

3.3 Understanding Leapfrog Urbanization Along Metropolitan Gradients

Urban sprawl research gained momentum in the 1980s and 1990s, when scholars began to recognize suburbs as autonomous entities essential to metropolitan growth. Los Angeles, an archetype of North American sprawl, saw its population density rise from approximately 4,600 inhabitants per square mile in 1950 to 6,000 by 1990, becoming the most populous 'suburban' area in the United States [4].

European suburbs have followed possibly different trajectories, shaped by distinct economic, social, and planning processes, yet display similar patterns of spatial dispersion [75]. In the Paris metropolitan area, the early 1980s saw a decline in inner-suburb population and density, accompanied by growth in peri-urban districts. Between 1960 and 1990, populations in outer suburbs and remote areas expanded steadily, primarily through migration from city centers and inner suburbs. By the late

Fig. 3.5 Residual green voids in a compact building tissue. *Source* authors' photographic archive

1990s, central Paris maintained exceptionally high densities, comparable to other compact Mediterranean cities. Over time, government interventions have led European suburban forms to increasingly resemble North American patterns, albeit with significant regional variation [29].

Suburbanization across Europe has unfolded unevenly, reflecting historical, economic, and cultural differences [76]. Northern European cities experienced earlier and more orderly suburban expansion [77], whereas southern European cities, particularly in Italy and Spain, underwent rapid but less structured decentralization (e.g. [78]). European suburbs generally exhibited low densities and a strong reliance on private transportation, functionally resembling the North American model [79].

Fig. 3.6 Urban sprawl, land take and vacant land with relict ecological value. *Source* authors' photographic archive

Empirical research should examine these dynamics more deeply, integrating historical perspectives with contemporary information to guide policy interventions that harmonize growth, equity, and environmental stewardship (Fig. 3.6). Suburbanization and urban decentralization definitely remain enduring features of contemporary metropolitan development [80]. Policies answering such challenge should balance residential needs, economic development, and sustainable land use in increasingly complex metropolitan environments [81].

3.3.1 Urban Morphology, Fringe Development and Vacant Land

While settlement expansion continues to be a central concern for contemporary cities, empirical research has predominantly focused on conceptualizing and modeling the mechanisms underlying urban growth, as well as developing analytical and composite metrics to evaluate its spatial outcomes [82]. Additionally, urban expansion was sometimes understood as an unplanned or semi-planned process characterized by spatially discontinuous settlements that encroach upon surrounding rural or

Fig. 3.7 Quasi-informal settlements and green voids in the 'vertical' city. *Source* authors' photographic archive

agricultural landscapes [83]. Such growth was frequently associated with negative environmental and social consequences (Fig. 3.7).

Urban expansion can be estimated adopting quantitative metrics and qualitative descriptors, each offering insights into different dimensions of development while presenting distinct limitations [84]. From a quantitative standpoint, density is a fundamental indicator, typically measured as the number of residents or employed individuals per square kilometer. Settlement density remains a defining feature of urban expansion and provides an effective tool for planning and economic assessment [85]. Continuity, as a morphological criterion, captures the extent to which land is urbanized in an uninterrupted manner. However, the presence of discontinuities does not inherently indicate undesirable growth, as low- or moderate-density urban centers may also exhibit discontinuous patterns [86]. Concentration reflects the spatial distribution of residential and non-residential units within a given area, indicating the degree of uniformity, while aggregation measures the extent to which development is clustered within a micro-area, minimizing land consumption [87]. Centrality evaluates the spatial relationship of an area to core urban districts, with low centrality frequently cited as a hallmark of dispersed urban growth—typically measured in kilometers from the city center.

Nuclearity refers to the presence or absence of a dominant urban center that consolidates multiple functions, or, conversely, to the existence of specialized centers for distinct activities. Identifying the primary center—whether residential or functional—and assessing the distribution of other significant nodes provides insights into the degree of functional dispersion within metropolitan landscapes [88]. Additional dimensions, including differentiated land use and proximity, describe the heterogeneity and spatial relationships of urban functions, highlighting the multifaceted nature of metropolitan expansion [89].

The notion of proximity specifically evaluates the spatial distance between distinct functions, such as the separation between residences and workplaces within a defined area. Many of these dimensions are closely linked to the spatial distribution of the population, as well as to typological and socio-spatial characteristics of the built environment. This observation underscores the need for interpretative frameworks that move beyond traditional planning approaches, which primarily focus on land-use patterns, and conventional economic analysis, which emphasize productive capacities and socio-demographic characteristics [90–93]. Various combinations of discontinuity, low centrality, and functional dispersion generate distinct forms of expansion, each reflecting specific historical, economic, and social contexts [94].

A comprehensive understanding of urban growth mechanisms necessitates integrating morphological, functional, and socioeconomic analysis (e.g. [95]). Recognizing these multiple dimensions allows planners and policymakers to identify e.g. areas most affected by uncontrolled development and to develop strategies aimed at mitigating environmental impacts, enhancing social cohesion, and promoting sustainable urban forms [96]. By framing urban growth as an evolving process rather than a static condition, it becomes possible to examine both its causes and consequences, providing a foundation for interventions that balance development, land conservation, and the equitable distribution of urban opportunities [97].

3.4 Settlement Diffusion, Economic Functions and Urban Voids

Functional differentiation is one of the most relevant organizing principles of metropolitan space, alongside mass consumption patterns. It can be quantified by comparing the average density of one type of use (e.g. residential) to that of another use within the same spatial unit. Additionally, functional differentiation may be assessed by considering the number and variety of functions—beyond residential—that are accessible from any point within the area, within a distance deemed reasonably walkable (for instance, 500 m, equivalent to approximately ten minutes on foot). Based on the diversity and accessibility of these functions, it becomes possible to identify the degree of functional integration within a given territory [72, 74, 98, 99].

Extensive scholarship has addressed processes of urban diffusion *vis à vis* functional differentiation in metropolitan regions, particularly in North American cities. Here, the phenomenon has been highly heterogeneous since the post-World War

II period, and even more so during the 1960s, due to distinctive planning decisions and residential models. American cities, especially at the turn of the nineteenth century, began to evolve according to patterns that diverged from the traditional European model, exhibiting a radio-centric form of urban growth and a kaleidoscope of functions being substantially decoupled from the distance to inner cities [60].

At the same time, research contributions from French-speaking, Germanic, and Mediterranean regions have increased substantially in recent years. [100] These studies try to underscore the importance of examining not only the spatial characteristics of expansion—basically at operational scales appropriate for assessing settlement and landscape features—but also the evolving relationship between urban form and functions in the old continent [101]. Purely functional approaches, in contrast, tend to equate the peri-urban area with commuting zones and areas of demographic revitalization—a perspective adopted e.g. in French geography, which has informed ongoing debates concerning the definition and governance of metropolitan systems (e.g. [102]).

Within this evolving mainstream, peri-urbanization has become a key subject of investigation, situated within the broader goal of understanding the economic and territorial reorganization of contemporary cities [103]. Scholarly attention in this phase increasingly focuses on systemic and functional approaches to urban networks, and the relational mechanisms that enable economic growth (Fig. 3.8). This includes examining the interplay between local and global dynamics in light of assessing the incipient urban globalization (e.g. [104]).

3.4.1 Sub-Central Expansion and Landscape Transformations

Landscape transformations redirect attention away from the traditional study of urban forms as a privileged scale for analyzing spatial texture, natural environments, social and economic systems, and the production of identities and socio-spatial conflicts [105]. Consequently, landscapes, settlement patterns, and principles of urban growth have often been regarded as of diminished interest, while the debate increasingly frames cities as nodes within global production networks. Considering urban expansion not only in terms of physical settlement but also in socioeconomic terms situates the phenomenon within the broader process of urban reorganization, highlighting its implications for development opportunities [106]. For instance, the so-called 'emerging areas' in Europe were interpreted during the 1990s as broadening the base of the urban hierarchy, with higher-order services concentrated in the largest cities (e.g. [107]). This did not imply a reversal of the traditional (metropolitan) hierarchy, but rather suggested a shift toward urban relations characterized by complementarity, specialization, and cooperative exchange [108].

The evolution of metropolitan areas toward a multipolar configuration reflects this shift. However, formation of multiple central nodes within a metropolitan region does

Fig. 3.8 Traditional olive groves under conversion to urban use. *Source* authors' photographic archive

not indicate a decline in the relevance of the city; rather, it corresponds to functional specialization, which can enhance the competitive potential of the entire area [109]. Decentralized hubs, often aligned with major transport nodes, may encourage more compact settlements around these centers, reducing the dispersive tendencies of urban growth [110]. Such configurations, sometimes described as 'non-diffusive branching' or decentralized concentration, illustrate urban expansion extending beyond compact city cores while maintaining relatively high densities in smaller settlements [111]. These settlements are strategically positioned to optimize accessibility to central areas, while internal road networks reduce dependence on private vehicles (Fig. 3.9).

This transition finally led metropolitan regions to depart from the classical center-periphery model, with the geography of marginality taking on complex, patchwork-like structures rather than simple concentric arrangements [112]. Contemporary cities increasingly consist of interconnected clusters and specialized nodes, with scale and mass sufficient for international competition, while retaining the potential for further expansion [113]. Even within such complex frameworks, urban expansion remains, however, a phenomenon mixing settlement and landscape transformations, characterized by diversified housing patterns, fragmented land use, and mixed urban-agricultural matrices [114]. Recognizing urban expansion in this way provides valuable insights for planning, enabling the identification of locational tendencies, critical growth elements, and coherent policy strategies [115].

Fig. 3.9 Mixed 'technocratic' landscapes with vacant land in sub-central locations. *Source* authors' photographic archive

3.5 Vacant Landscapes In-Between Settlement Morphology and Metropolitan Structure

Investigations carried out at broader territorial scales often simplify the analysis of landscape elements [116]. On the one hand, landscape ecologists, many regional economists, and the majority of urban planners advocate interventions aimed at containing urbanization to promote sustainable development and reduce land consumption [117]. On the other hand, geographers, some sociologists, and planners approach urban expansion as an opportunity to interpret emerging mechanisms of territorial organization, seeking to identify both the potential benefits and risks of growth without compromising local identities, endogenous capacities, or environmental values [118].

From a settlement perspective, vacant land can be exemplified as areas of rarefied built density interspersed with green, agricultural, and semi-natural spaces, where diverse urban elements—shopping centers, industrial facilities, single- and multi-family residences, and transport infrastructures—intermingle. These areas often convey a sense of spatial inhomogeneity and transition, combining new subdivisions with pre-existing rural activities. The phenomenon exhibits multidimensionality, encompassing growth rates, density, spatial configuration, accessibility, and qualitative or aesthetic dimensions (e.g. [119]). The associated urban expansion is characterized by discontinuous development, low-density residential areas typically composed of single-family homes with private gardens, linear commercial developments along major transport routes, functional specialization with spatial segregation, and limited accessibility that increases reliance on private vehicles [120].

Moving from residential to industrial land use, commercial strips—commonly located along arterial roads and motorway exits—reflect strategic planning for accessibility, parking, and visibility, resulting in minimal landscaping and fragmented public space. Buildings are generally single-story and spacious, while differentiated land uses are frequently implemented through zoning instruments, emphasizing functional segregation [121]. In such contexts, the meaning and configuration of urban expansion vary across spatial scales, reflecting local territorial contexts, density gradients, connectivity, and diversity of land use influencing size, shape and functions of vacant land [122].

3.5.1 Urban Growth, Vacant Land and Settlement Aestethic

Urban expansion also generates a distinctive and novel urban aesthetic, characterized by fragmentation, juxtaposition, and a multiplicity of forms and styles. We have previously mentioned how city regions no longer present the coherent 'formal' logic of compact, traditional urban cores, but instead manifests as a dispersed, discontinuous assemblage of architectural episodes. This aesthetic may be described as a sort of 'acrasia', where the juxtaposition of disparate visual elements produces a complex, non-hierarchical urban image [123]. Whereas modernist urban design aimed at creating universal forms—a 'house for all', postmodern metropolitan expansion celebrates individualized settlement expressions. This is often realized through mass-customized housing, hyperpersonalized consumer spaces, and leisure facilities [124]. Private residences, shopping malls, office parks, and recreational complexes increasingly reflect personal, corporate, or commercial identities, producing a hyperreal landscape in which stylistic coherence is replaced by visual heterogeneity [36].

The visual footprint of such territories is marked by abrupt shifts in scale, style, and materiality (Fig. 3.10). Navigating dispersed urban areas, one encounters a mosaic of images: oversized commercial signage, residential buildings with kitsch or eclectic ornamentation, industrial structures, leisure complexes, and retail outlets deliberately designed to evoke historical or fantastical scenarios. This fragmented landscape communicates a sense of disorder, yet it also reflects broader social and economic dynamics, where consumer choice, corporate branding, and private ownership shape the aesthetic outcome. From the perspective of landscape aesthetics, such heterogeneity challenges traditional notions of harmony, proportion, and continuity [125]. Rather than privileging unity and coherence, this structure embraces variety, discontinuity, and aesthetic pluralism, producing territories where visual and functional organization is both decentralized and unevenly scaled.

Fig. 3.10 The balanced aesthetic of urban panoramas with a green void. *Source* authors' photographic archive

At the architectural scale, this new aesthetic often favors self-referential forms that disregard their immediate context. Star-architect designs, such as Frank Gehry's Guggenheim Museum in Bilbao, function as iconic sculptures, visually and conceptually independent from surrounding structures and landscapes. These buildings operate as instruments of urban marketing, drawing attention through novelty, monumentality, and aesthetic spectacle rather than contextual integration. Rem Koolhaas theorizes this phenomenon as 'junkspaces' and the 'generic city': environments that expand continuously without cohesive planning, characterized by eclectic forms, sites chosen for market convenience rather than urban logic, and densities lower than traditional urban cores. From an urban design perspective, these territories are organized around infrastructure corridors or simply available land, rather than socially or aesthetically meaningful hierarchies, resulting in sprawling networks of autonomous architectural fragments. In some instances, they could even be regarded as 'implicit voids' when confronted with the surrounding landscape matrix.

Public space within these expanded territories also undergoes profound transformation. Traditional squares, parks, and pedestrian streets are often absent or marginalized, replaced by privatized pseudo-public environments such as shopping malls and leisure complexes—possibly intended as a latent form of urban void. These spaces mimic the social functions of conventional public areas but are controlled, commodified, and oriented toward consumption rather than civic engagement. Streets within these developments may resemble film sets, designed to simulate urban interaction while prioritizing vehicle circulation, visual spectacle, or commercial accessibility. The landscape of such edge-cities is defined less by the formal qualities of traditional urbanism—proportion, axiality, visual coherence—and more by experiential qualities: sequence, surprise, and the tension between continuity and rupture.

Landscape aesthetics further illuminates the consequences of dispersed urban expansion. The interplay between built elements, open spaces, and natural features is often incidental rather than deliberate. Agricultural patches, river corridors, and remnant green spaces are interspersed with residential and commercial development, generating a mosaic of contrasting textures, colors, and densities. This type of landscape challenges conventional aesthetic judgments, inviting a recalibration of criteria for beauty, legibility, and coherence in urban design.

Based on these premises, the aesthetics of contemporary urban expansion emphasizes fragmentation, eclecticism, and hyper-individualization. The visual and spatial heterogeneity of these territories reflects broader socioeconomic processes, including consumerism, privatization, and market-driven land allocation. While the resulting landscape may appear disorderly when measured against traditional urban aesthetics, it offers new forms of spatial experience, combining mobility, surprise, and a multiplicity of visual and functional cues. Understanding these phenomena from the perspectives of landscape aesthetics is essential for planners and designers seeking to guide contemporary urban growth in ways that are visually engaging, socially meaningful, and ecologically sensitive. By integrating considerations of form, function, and perception, design of both empty and void landscapes can navigate the challenges of dispersed development while fostering environments that balance individual expression with collective legibility and environmental quality.

3.6 Peri-Urban Expansion, Spatial Planning, and Vacant Land

Contemporary urban expansion finally represents a profound paradigm shift in planning practices, reflecting the increasing influence of market forces over the development of post-metropolitan regions. Political and administrative authorities, which once exercised significant control during the so-called 'golden years of planning' associated with modernist urbanism, have seen their roles diminish, particularly under the influence of neoliberal policies from the 1980s onward. This transition has marked the decline of large-scale, coordinated planning programs that were traditionally implemented by European states under welfare-oriented logics.

Today, urban growth is often driven by private initiative, operating within a framework of deregulation and characterized by fragmented, uncoordinated interventions. Such market-driven expansion sharply contrasts with the systematic and functionalist planning principles of the modern city, where urban development was closely aligned with industrial, social, and infrastructural objectives. Municipal administrations increasingly adopt reactive strategies, retrofitting infrastructure to accommodate ongoing private development, while the boundaries and limits of urban expansion are shaped less by intentional design than by indirect constraints, such as topography, existing transport networks, and the availability of services.

From the inherent perspective of land policy, exurban development can be interpreted as the outcome of a profound paradigm shift in spatial planning. Market forces have come to play a central role in shaping post-metropolitan territories, while political and administrative bodies—once dominant during the so-called 'golden years' of modern spatial planning—have seen their authority significantly reduced, particularly under the neoliberal wave of the 1980s. This transition marked the decline of large-scale, unitary planning carried out e.g. by European states under welfare-oriented logics. Private actors now drive urban development in mostly deregulated contexts, engaging in fragmented, uncoordinated, and often ad hoc interventions. Such market-driven expansion contrasts sharply with the functionalist principles and Fordist industrial logic that underpinned modern spatial planning. Local administrations frequently find themselves reacting to private initiatives, retrofitting infrastructure to accommodate ongoing growth rather than shaping it through proactive design (e.g. [126]).

This interplay between *laissez-faire* policies and residual planning frameworks has produced horizontally structured (and sometimes unintendedly polycentric) territories, in which urban artifacts (and landscape voids) appear arranged according to principles of apparent randomness. The resulting phenomenon could be described as the 'amoeba city', with Los Angeles again cited as the archetype of dispersed, postmodern urbanism. The perceived disorder and randomness of fringe landscapes mixing built-up and vacant spaces derive from the absence of clearly defined edges, and the emergence of unprecedented proximities between residential, commercial, and industrial functions. Within these territories, new social and functional mixes replace those of the traditional compact city, redefining centralities and marginalities [127].

The elusiveness intrinsic in these territories lies in the fact that they cannot be readily apprehended in their entirety. For instance, several of Europe's largest conurbations now cover territories approaching the scale of entire countries, such as the network connecting Amsterdam, Utrecht, Rotterdam, and The Hague, or the Ruhr region. This has led spatial planners to increasingly adopt the notion of 'city-region' to describe these complex, composite spaces. Such territories incorporate historical urban cores, metropolitan centers, peri-urban landscapes, industrial areas, vacant land, and office complexes, creating urban fabrics that are simultaneously continuous, fragmented, and hybrid. Settlement patterns may follow linear corridors, circular arrangements, or disperse from major infrastructural axes, forming expansive meshes with irregular, frayed qualities. Growth within these territories is rarely isotropic, as physical, geomorphological, historical, and social constraints interact with axes of accessibility to facilitate or restrict expansion along specific directions.

Peri-urban areas, however, still demonstrate self-organizing dynamics that may reveal mechanisms of sustainable development. Beyond purely economic considerations, such areas highlight the role of quality of life, social cohesion, and identity formation in shaping metropolitan growth [128]. Even within this critical and innovative perspective, public institutions and decision-makers remain central, translating analytical knowledge into concrete territorial interventions. In this context, robust analytical tools are essential to support planning decisions amid uncertainty, integrating economic development goals with emerging social, environmental, and infrastructural priorities [129].

3.6.1 Planning New Forms of Contemporary Cities Under a 'Land Take' Era

We have previously argued how urban expansion can be understood not only as a process of economic development but also, and perhaps more importantly, as a distinct morphological phenomenon that shapes the spatial structure of contemporary cities. These decisions determine the arrangement of residential, commercial, industrial, recreational, and vacant spaces, generating complex urban fabrics whose characteristics are influenced by density, functional diversity, and the capacity of transport networks. In particular, the spatial distribution of land uses and the accessibility provided by transportation infrastructure are critical in defining the functional distances between places where people live, work, and engage in leisure activities. High-density, mixed-use areas promote shorter travel distances, whereas low-density, functionally homogeneous developments increase travel distances, foster dependence on private vehicles, and often result in inefficient land use—possibly creating voids—and heightened environmental impacts.

In this perspective, land consumption has become a major challenge for both established and emerging metropolitan regions. This phenomenon manifests in several ways, including the physical occupation of high-quality agricultural or ecologically valuable soils, as well as the allocation of soils for low-value or economically marginal purposes—possibly consolidating vacant land. Greenfield urban expansion not only depletes ecological functions and ecosystem services but also erodes the cultural, social, and economic value of these spaces. Open and green areas contribute to visual quality, recreational opportunities, local identity, and even economic benefits such as tourism, highlighting their multifaceted importance in sustaining livable and resilient environments [130].

In response to these challenges, spatial planners have developed a variety of techniques aimed at containing urban expansion and protecting valuable land. Downzoning reduces the permissible density of development in designated areas, maintaining separation between buildings. However, it often perpetuates low-density, car-dependent patterns without enhancing the provision of public space. Clustering concentrates development within specific zones, leaving surrounding land available for agriculture, forestry, or public recreation, thereby balancing urban growth with conservation objectives [131]. Agricultural zoning specifically protects farmland, sustains local economies, and preserves community characteristics by restricting development, even for private purposes (Fig. 3.11). Growth barriers, particularly in the form of urban boundaries, are designed to contain development within defined limits, creating green belts that protect natural landscapes while encouraging higher-density infill within the urban core [132]. These measures can be implemented at local or regional scales: broader implementation increases overall effectiveness, whereas local adoption allows greater flexibility and citizen participation in planning processes [133].

The application of growth management techniques also produces significant socioeconomic effects. By limiting buildable land, growth barriers can increase property values within the urban core, incentivizing higher-density construction, mixed-use developments, and innovative solutions such as underground parking. These mechanisms promote more efficient land use, reduce spatial segregation, and help maintain the social and economic vitality of city centers. The experience of Portland, Oregon, illustrates the potential of integrated planning approaches to guide urban expansion in ways that balance growth with the preservation of natural and agricultural land, fostering sustainable and resilient metropolitan regions [134].

Fig. 3.11 Urban farming as a way of managing fertile soils and green voids. *Source* authors' photographic archive

Beyond land management techniques, sustainable urban growth claims for a broader understanding of the interdependencies between density, functional diversity, mobility, and environmental stewardship. Vertical and horizontal differentiation of land uses enhances accessibility, encourages walking and cycling, and reduces reliance on private vehicles. The spatial organization of functions influences routine travel patterns: greater density and land-use diversity are correlated with shorter trips and higher overall efficiency. Simultaneously, preserving open spaces and green belts supports environmental, social, and economic resilience, sustaining biodiversity, recreational opportunities, and community well-being [135].

Effective land management definitely requires coordinated strategies that align public policy, market forces, and community priorities, ensuring that growth is spatially coherent, socially inclusive, and ecologically sustainable. By reconciling the pressures of private development with the need to preserve natural and agricultural land, urban planners can guide cities toward forms of expansion that are both resilient and adaptable, capable of supporting the diverse needs of contemporary populations while safeguarding the integrity of the natural and built environment for future generations.

3.7 Concluding Remarks

If the logic of the compact city was historically based on a dense and coherent urban fabric, in which buildings were arranged around streets, squares, and open spaces that constituted the civic and social heart of the city, contemporary metropolitan expansion presents a markedly different scenario. In these new settlements, public space often assumes a residual or incidental character, appearing more as a tabula rasa than as a lively arena for social interaction. Unlike the neighborhoods described by Jane Jacobs, where local commerce, pedestrian activity, and social encounters structured the urban experience, the public spaces of contemporary cities are frequently unfinished, loosely defined, and insufficiently habitable. The civic dimension that once animated streets and squares has, in many cases, been supplanted by an individualistic ethos, in which private interests and economic considerations dictate spatial organization. Courtyards, shopping complexes, and gated enclaves replace communal areas, fostering isolated residential patterns that weaken the traditional social bonds of urban life.

The aesthetic consequences of this transformation are significant. Whereas the compact city offered a coherent visual and functional hierarchy, contemporary urban expansion produces fragmented (and visually vacant) landscapes defined by heterogeneity and discontinuity. Streets, open areas, and architectural elements are no longer linked by a unifying formal logic, but are juxtaposed according to private development agendas and market forces. This results in a visual landscape marked by abrupt shifts in scale, materiality, and stylistic language, reflecting a broader postmodern sensibility of pluralism and eclecticism. From the perspective of landscape aesthetics, these territories challenge conventional notions of harmony, proportion, and legibility. Their aesthetic value arises not from unity or coherence, but from the dynamic interplay of fragments, the tension between continuity and rupture, and the experiential sequencing of movement through heterogeneous spaces.

Despite these transformations, European cities often retain the vitality and symbolic function of their historic public spaces. Carefully preserved squares and streets continue to anchor social life, supported by aesthetic interventions and urban marketing strategies that enhance both visual and experiential appeal. In contrast, in North American cities and many South American megalopolises, public space is largely privatized and insularized, organized around class, lifestyle, or consumer preference. Centralized planning and collective urban interventions have historically been perceived as intrusions on private autonomy, reinforcing a cultural and ideological commitment to suburban escape. The single-family home, situated within a private lot and maintained as a personal garden, embodies this ideal, fostering residential decentralization and creating hybrid landscapes that oscillate between the urban and pastoral—neither fully consolidated nor purely bucolic.

The concept of refuge, or heterotopia, is central to understanding the aesthetic and functional qualities of contemporary urban expansion. Drawing on M. Augé's reflections on 'non-places' including shopping malls and thematic parks, these territories emerge as void islands within a network of experiences, each offering controlled and curated environments for consumption, leisure, and social interaction. Residents navigate this landscape through a form of experiential 'zapping', moving fluidly between work, home, and recreation, facilitated by mechanized transport and the compression of space–time dimensions. The resulting landscape is characterized by hybridization, decentralization, fragmentation, and a nebulous spatial organization, challenging traditional epistemological frameworks and demanding new paradigms for urban analysis, planning, and design.

Local context remains a decisive factor in shaping the specific forms and qualities of urban expansion. While certain characteristics—such as low density, fragmentation, and privatized public space—are recurrent, their expression varies according to economic, social, and cultural conditions. The urban expansion of the Po Valley in Northern Italy, with its dense network of services, embedded small and medium-sized enterprises, and strong civic culture—differs markedly from the sprawling, often informally constructed landscapes of Southern regions, or from Mediterranean coastal areas transformed by intensive tourism. These variations underscore the need to understand urban expansion not as a singular model, but as a heterogeneous phenomenon mediated by local history, policy, economy, and social practice.

In terms of aesthetics and urban experience, these territories demand new interpretive approaches. Traditional frameworks that valorize visual coherence, axiality, and hierarchical organization are insufficient for understanding the complexity of fragmented and decentralized landscapes. Instead, aesthetic evaluation should account for sequence, contrast, and relational perception, emphasizing how movement through space, temporal experience, and the juxtaposition of visual and functional elements—including both 'empty' and 'void' land—shape the lived environment. This perspective highlights both challenges and potential of contemporary urban expansion: while the civic and formal cohesion of traditional cities may be attenuated, new forms of visual and social experience emerge, offering alternative paradigms for inhabitation, identity, and engagement with urban fabrics. Hybrid landscapes, heterotopic spaces, and dispersed morphologies thus occupy a central role in twenty-first century urban studies, requiring analytical, design, and policy frameworks capable of addressing their complexity while enhancing both aesthetic quality and social vitality.

References

1. Angel S, Parent J, Civco DL, Blei A, Potere D (2011) The dimensions of global urban expansion: estimates and projections for all countries, 2000–2050. Prog Plan 75(2):53–107
2. Brenner N, Schmid C (2014) The 'urban age' in question. Int J Urban Reg Res 38(3):731–755
3. Davoudi S (2003) European briefing: polycentricity in European spatial planning: from an analytical tool to a normative agenda. Eur Plan Stud 11(8):979–999
4. Bruegmann R (2019) Sprawl: a compact history. University of Chicago Press, Chicago
5. Beriatos E, Gospodini A (2004) "Glocalising" urban landscapes: Athens and the 2004 Olympics. Cities 21:187–202
6. Di Feliciantonio C, Salvati L, Sarantakou E, Rontos K (2018) Class diversification, economic growth and urban sprawl: Evidences from a pre-crisis European city. Qual Quant 52:1501–1522
7. Rivolin UJ, Faludi AKF (2005) The hidden face of European spatial planning. Eur Plan Stud 13(2):195–215
8. Power A (2001) Social exclusion and urban sprawl: is the rescue of cities possible? Reg Stud 35(8):731–742
9. Paul V, Tonts M (2005) Containing urban sprawl: trends in land use and spatial planning in the metropolitan region of Barcelona. J Environ Planning Manage 48(1):7–35
10. Nickayin SS, Tomao A, Quaranta G, Salvati L, Gimenez Morera A (2020) Going toward resilience? Town planning, peri-urban landscapes, and the expansion of Athens Greece. Sustainability 12(24):10471
11. Kasanko M, Barredo JI, Lavalle C, McCormick N, Demicheli L, Sagris V, Brezger A (2006) Are European cities becoming dispersed? a comparative analysis of fifteen european urban areas. Landsc Urban Plan 77(1–2):111–130
12. Kahn ME (2000) The environmental impact of suburbanization. J Policy Anal Manage 19(4):569–586
13. Halbac-Cotoara-Zamfir R, Marucci A, Salvia R, Quaranta G, Sateriano A, Cecchini M, Bianchini L (2022) Caring of the fringe? Mediterranean desertification between peri-urban ecology and socioeconomics. Sustainability 14(3):1426
14. Genske DD (2003) Urban land—degradation, investigation, remediation. Springer, Berlin
15. Galster G, Hanson R, Ratcliffe MR, Wolman H, Coleman S, Freihage J (2001) Wrestling sprawl to the ground: defining and measuring an elusive concept. Hous Policy Debate 12(4):681–717
16. Ioannidis C, Psaltis C, Potsiou C (2009) Towards a strategy for control of suburban informal buildings through automatic change detection. Comput Environ Urban Syst 33:64–74
17. Homsy GC, Warner ME (2015) Cities and sustainability: polycentric action and multilevel governance. Urban Affairs Review 51(1):46–73
18. Domene E, Sauri D (2006) Urbanization and water consumption: influential factors in the metropolitan region of Barcelona. Urban Stud 43(9):1605–1623
19. Keddie J, Tonkiss F (2010) The market and the plan: housing, urban renewal and socio-economic change in London. City Cult Soc 1(2):57–67
20. Domene E, Sauri D, Parés M (2005) Urbanization and sustainable resource use: the case of garden watering in the metropolitan region of Barcelona. Urban Geogr 26(6):520–535

21. Herrschel T (2009) City regions, polycentricity and the construction of peripheralities through governance. Urban Res Pract 2(3):240–250
22. Kemeny T, Storper M (2012) The sources of urban development: wage, housing, and amenity gaps across American cities. J Reg Sci 52(1):85–108
23. Pacione M (2003) Urban environmental quality and human wellbeing—a social geographical perspective. Landsc Urban Plan 65:19–30
24. Masini E, Tomao A, Barbati A, Corona P, Serra P, Salvati L (2019) Urban growth, land-use efficiency and local socioeconomic context: a comparative analysis of 417 metropolitan regions in Europe. Environ Manage 63:322–337
25. Kroll F, Kabisch N (2012) The relation of diverging urban growth processes and demographic change along an urban-rural gradient. Popul Space Place 18:260–276
26. Mudu P (2006) Patterns of segregation in contemporary Rome. Urban Geogr 27(5):422–440
27. Parés M, March H, Saurí D (2013) Atlantic gardens in mediterranean climates: understanding the production of suburban natures in Barcelona. Int J Urban Reg Res 37:328–347
28. Mouratidis K (2019) Compact city, urban sprawl, and subjective well-being. Cities 92:261–272
29. Richardson H, Chang-Hee C (2004) Urban sprawl in Western Europe and the United States. Ashgate, London
30. King R, Proudfoot L, Smith B (1997) The Mediterranean. Environment and society, Arnold, London
31. Orenstein DE, Frenkel A, Jahshan F (2014) Methodology matters: measuring urban spatial development using alternative methods. Environ Plann B Plann Des 41:3–23
32. Vinci S, Egidi G, López Gay A, Salvati L (2022) Population growth and urban management in metropolitan regions: the contribution of natural balance and migration to polycentric development in Barcelona. Appl Spat Anal Policy 15(1):71–94
33. Tomao A, Quatrini V, Corona P, Ferrara A, Lafortezza R, Salvati L (2017) Resilient landscapes in Mediterranean urban areas: Understanding factors influencing forest trends. Environ Res 156:1–9
34. Scott AJ, Storper M (2015) The nature of cities: the scope and limits of urban theory. Int J Urban Reg Res 39(1):1–15
35. Koprowska K, Łaszkiewicz E, Kronenberg J (2020) Is urban sprawl linked to green space availability? Ecol Ind 108:105723
36. Kloosterman RC, Musterd S (2001) The polycentric urban region: towards a research agenda. Urban Studies 38(4):623–633
37. Lee S (2011) Metropolitan growth patterns and socio-economic disparity in six US Metropolitan areas 1970–2000. Int J Urban Reg Res 35(5):988–1011
38. Liu L, Meng L (2020) Patterns of urban sprawl from a global perspective. J Urban Plan Dev 146(2):04020004
39. Longhi C, Musolesi A (2007) European cities in the process of economic integration: towards structural convergence. Ann RegNal Sci 41:333–351
40. Makse HA, Andrade JS, Batty M, Havlin S, Stanley HE (1998) Modeling urban growth patterns with correlated percolation. Phys Rev E 58(6):7054
41. Le Goix R (2005) Gated communities: sprawl and social segregation in Southern California. Hous Stud 20(2):323–343
42. Lindstrom MJ, Bartling H (2003) Suburban sprawl: culture, theory, and politics. Rowman & Littlefield, New York
43. Musterd S (2005) Social and ethnic segregation in Europe: levels, causes and effects. J Urban Aff 27(3):331–348
44. Ismael HM (2021) Urban form study: the sprawling city—review of methods of studying urban sprawl. GeoJ 86(4):1785–1796
45. Feng Y, Wang X, Du W, Liu J, Li Y (2019) Spatiotemporal characteristics and driving forces of urban sprawl in China during 2003–2017. J Clean Prod 241:118061
46. Egidi G, Salvati L, Vinci S (2020) The long way to tipperary: City size and worldwide urban population trends, 1950–2030. Sustain Cities Soc 60:102148

47. Allen J, Barlow J, Maloutas T, Padovan L (2004) Housing in Southern Europe. Blackwell, London
48. Andreotti A, Le Galès P, Moreno Fuentes FJ (2013) Controlling the urban fabric: the complex game of distance and proximity in European upper-middle-class residential strategies. Int J Urban Reg Res 37(2):576–597
49. Antrop M (2004) Landscape change and the urbanization process in Europe. Landsc Urban Plan 67:9–26
50. Arapoglou VP, Sayas J (2009) New facets of urban segregation in Southern Europe. Eur Urban Reg Stud 16(4):345–362
51. Arbaci S (2008) (Re) viewing ethnic residential segregation in southern European cities: housing and urban regimes as mechanisms of marginalisation. Hous Stud 23(4):589–613
52. Arbaci S, Malheiros J (2010) De-segregation, peripheralisation and the social exclusion of immigrants: Southern European Cities in the 1990s. J Ethn Migr Stud 36(2):227–255
53. Couch C, Petschel-held G, Leontidou L (2007) Urban sprawl in Europe: landscapes, land-use change and policy. Blackwell, London
54. Di Feliciantonio C, Salvati L (2015) 'Southern' alternatives of urban diffusion: investigating settlement characteristics and socio-economic patterns in three mediterranean regions. Tijdschr Econ Soc Geogr 106(4):453–470
55. Buzar S, Ogden P, Hall R, Haase A, Kabisch S, Steinfiihrer A (2007) Splintering urban populations: emergent landscapes of reurbanisation in four European Cities. Urban Stud 44(4):651–677
56. Colantoni A, Grigoriadis E, Sateriano A, Sarantakou E, Salvati L (2017) Back to Von Thunen: a Southern European perspective on mono-centric urban growth, economic structure and non-urban land decline. Int Plan Stud 22(3):173–188
57. Chorianopoulos I, Pagonis T, Koukoulas S, Drymoniti S (2010) Planning, competitiveness and sprawl in the Mediterranean city: the case of Athens. Cities 27:249–259
58. Cassiers T, Kesteloot C (2012) Socio-spatial inequalities and social cohesion in European Cities. Urban Stud 49(9):1909–1924
59. Gospodini A (2006) Portraying, classifying and understanding the emerging landscapes in the post-industrial city. Cities 23(5):311–330
60. Johnson DL, Lewis LA (2007) Land degradation—creation and destruction. Rowman & Littlefield, Lahnam
61. Gkartzios M (2013) 'Leaving Athens': narratives of counterurbanisation in times of crisis. J Rural Stud 32:158–167
62. De Rosa S, Salvati L (2016) Beyond a 'side street story'? Naples from spontaneous centrality to entropic polycentricism, towards a 'crisis city.' Cities 51:74–83
63. Cuadrado-Ciuraneta S, Durà-Guimerà A, Salvati L (2017) Not only tourism: unravelling suburbanization, second-home expansion and "rural" sprawl in Catalonia. Spain. Urban Geography 38(1):66–89
64. Gargiulo Morelli V, Rontos K, Salvati L (2014) Between suburbanisation and re-urbanisation? revisiting the urban life cycle in a mediterranean compact city. Urban Research and Practice 7(1):74–88
65. Gkartzios M, Scott K (2015) A cultural panic in the province? Counterurban mobilities, creativity, and crisis in Greece. Popul Space Place 21(8):843–855
66. Faludi A (2000) The European spatial development perspective—what next? Eur Plan Stud 8(2):237–250
67. Faludi AKF (2006) From European spatial development to territorial cohesion policy. Reg Stud 40(6):667–678
68. Maloutas T (2004) Segregation and residential mobility. Socially entrapped social mobility and its impact on segregation in Athens. Eur Urban RegNal Stud 11(2):171–187
69. Maloutas T (2007) Socio-economic classification models and contextual difference: the "European socio-economic classes" (ESeC) from a South European angle. South Eur Soc Polit 12(4):443–460

70. Musterd S, Marcińczak S, Van Ham M, Tammaru T (2017) Socioeconomic segregation in European capital cities. Increasing separation between poor and rich. Urban Geogr 38(7):1062–1083
71. Delladetsima PM (2006) The emerging property development pattern in Greece and its impact on spatial development. Eur Urban RegNal Stud 13(3):245–278
72. Gospodini A (2009) Post-industrial trajectories of Mediterranean European cities: the case of post-Olympics Athens. Urban Studies 46:1157–1186
73. Grekousis G, Manetos P, Photis YN (2013) Modeling urban evolution using neural networks, fuzzy logic and GIS: the case of the Athens metropolitan area. Cities 30:193–203
74. Salvati L, Zitti M, Sateriano A (2013) Changes in city vertical profile as an indicator of sprawl: evidence from a mediterranean urban region. Habitat Int 38:119–125
75. Salvati L, Venanzoni G, Serra P, Carlucci M (2016) Scattered or polycentric? untangling urban growth in three southern European metropolitan regions through exploratory spatial data analysis. Ann Reg Sci 57:1–29
76. Turok I, Mykhnenko V (2007) The trajectories of European cities, 1960–2005. Cities 24(3):165–182
77. Storper M, Scott AJ (2016) Current debate in urban theory: a critical assessment. Urban Studies 53(6):1114–1136
78. Salvati L, Carlucci M (2016) Patterns of sprawl: the socioeconomic and territorial profile of dispersed urban areas in Italy. Reg Stud 50(8):1346–1359
79. Rubiera-Morollón F, Garrido-Yserte R (2020) Recent literature about urban sprawl: a renewed relevance of the phenomenon from the perspective of environmental sustainability. Sustainability 12(16):6551
80. Sutton PC, Goetz AR, Fildes S, Forster C, Ghosh T (2010) Darkness on the edge of town: mapping urban and peri-urban Australia using nighttime satellite imagery. Prof Geogr 62(1):119–133
81. Townsend A (2009) Integration of economic and spatial planning across scales. Int J Public Sect Manag 22(7):643–659
82. Vinci S, Vardopoulos I, Salvati L (2023) A tale of a shrinking City? exploring the complex interplay of socio-demographic dynamics in the recent development of Attica Greece. Cities 132:104089
83. Zambon I, Colantoni A, Salvati L (2019) Horizontal vs vertical growth: understanding latent patterns of urban expansion in large metropolitan regions. Sci Total Environ 654:778–785
84. Weber C, Petropoulou C, Hirsch J (2005) Urban development in the Athens metropolitan area using remote sensing data with supervised analysis and GIS. Int J Remote Sens 26(4):785–796
85. Salvati L, Sateriano A, Grigoriadis E, Carlucci M (2017) New wine in old bottles: the (changing) socioeconomic attributes of sprawl during building boom and stagnation. Ecol Econ 131:361–372
86. Salvati L, Ferrara A, Chelli F (2018) Long-term growth and metropolitan spatial structures: an analysis of factors influencing urban patch size under different economic cycles. Geogr Tidsskr-Dan J Geogr 118(1):56–71
87. Schneider A, Woodcock CE (2008) Compact, dispersed, fragmented, extensive? a comparison of urban growth in twenty-five global cities using remotely sensed data, pattern metrics and census information. Urban Studies 45(3):659–692
88. Vidal M, Domene E, Saurí D (2011) Changing geographies of water-related consumption: residential swimming pools in suburban Barcelona. Area 43(1):67–75
89. Seifollahi-Aghmiuni S, Kalantari Z, Egidi G, Gaburova L, Salvati L (2022) Urbanisation-driven land degradation and socioeconomic challenges in peri-urban areas: insights from Southern Europe. Ambio 51(6):1446–1458
90. Salvati L (2014) Towards a polycentric region? the socioeconomic trajectory of Rome, an 'eternal' Mediterranean city. Tijdschr Econ Soc Geogr 105(3):268–284
91. Soliman AM (2004) Regional planning scenarios in South Lebanon: the challenge of rural-urban interactions in the era of liberation and globalization. Habitat Int 28:385–408

92. Song X, Feng Q, Xia F, Li X, Scheffran J (2021) Impacts of changing urban land-use structure on sustainable city growth in China: a population-density dynamics perspective. Habitat Int 107:102296
93. Tsilimigkas G, Kizos T (2014) Space, pressures and the management of the Greek landscape. Geogr Ann: Ser B, Hum Geogr 96(2):159–175
94. Leontidou L (1996) Alternatives to modernism in (Southern) urban theory: exploring in-between spaces. Int J Urban Reg Res 20(2):180–197
95. Kourliouros E (1997) Planning industrial location in Greater Athens: the interaction between deindustrialization and anti-industrialism during the 1980s. Eur Plan Stud 5(4):435–460
96. Meijers E (2008) Measuring polycentricity and its promises. Eur Plan Stud 16(9):1313–1323
97. Sassen S (2010) The city: its return as a lens for social theory. City Cult Soc 1(1):3–11
98. Durà-Guimera A (2003) Population deconcentration and social restructuring in Barcelona, a European Mediterranean city. Cities 20(6):387–394
99. Leontidou L (1990) The Mediterranean city in transition. Cambridge University Press, Cambridge
100. Hasse JE, Lathrop RG (2003) Land resource impact indicators of urban sprawl. Appl Geogr 23:159–175
101. Giannakourou G (2005) Transforming spatial planning policy in Mediterranean countries: Europeanization and domestic change. Eur Plan Stud 13:319–331
102. Garcia-López MÀ (2019) All roads lead to Rome… and to sprawl? evidence from European cities. Reg Sci Urban Econ 79:103467
103. European Environment Agency (2006) Urban sprawl in Europe—the ignored challenge. EEA Report no. 10, Copenhagen
104. Serra P, Vera A, Tulla AF, Salvati L (2014) Beyond urban-rural dichotomy: exploring socioeconomic and land-use processes of change in Spain (1991–2011). Appl Geogr 55:71–81
105. Seto KC, Ramankutty N (2016) Hidden linkages between urbanization and food systems. Science 352(6288):943–945
106. Bianchini L, Egidi G, Alhuseen A, Sateriano A, Cividino S, Clemente M, Imbrenda V (2021) Toward a dualistic growth? Population increase and land-use change in Rome. Italy. Land 10(7):749
107. Alphan H (2003) Land use change and urbanisation of Adana, Turkey. Land Degrad Dev 14:575–586
108. Alexandri G (2015) Reading between the lines: gentrification tendencies and issues of urban fear in the midst of Athens' crisis. Urban Stud 52(9):1631–1646
109. Abozeid ASM, AboElatta TA (2021) Polycentric vs monocentric urban structure contribution to national development. J Eng Appl Sci 68(1):11
110. Marull J, Pino J, Tello E (2008) The loss of territorial efficiency: an ecological analysis of land-use changes in western Mediterranean agriculture. Global Environment 2:112–150
111. De Muro P, Monni S, Tridico P (2011) Knowledge-based economy and social exclusion: shadow and light in the roman socio-economic model. Int J Urban Reg Res 35(6):1212–1238
112. Pili S, Grigoriadis E, Carlucci M, Clemente M, Salvati L (2017) Towards sustainable growth? a multi-criteria assessment of (Changing) urban forms. Ecol Ind 76:71–80
113. Bhatta B (2010) Causes and consequences of urban growth and sprawl. Analysis of urban growth and sprawl from remote sensing data. Springer, Berlin
114. Zambon I, Serra P, Sauri D, Carlucci M, Salvati L (2017) Beyond the 'Mediterranean city': Socioeconomic disparities and urban sprawl in three Southern European cities. Geogr Ann: Ser B, Hum Geogr 99(3):319–337
115. Carlucci M, Grigoriadis E, Rontos K, Salvati L (2017) Revisiting an hegemonic concept: long-term 'Mediterranean Urbanization' in between city re-polarization and metropolitan decline. Appl Spat Anal Policy 10(3):347–362
116. Cakir G, Un C, Baskent EZ, Kose S, Sivrikaya F, Keles S (2008) Evaluating urbanization, fragmentation and land use/cover change pattern in Istanbul city, Turkey from 1971 to 2002. Land Degrad Dev 19:663–675

117. Brunori G, Rossi A (2007) Differentiating countryside. Social representations and governance patterns in rural areas with high social density: the case of Chianti. Italy. Journal of Rural Studies 23:183–205
118. Chorianopoulos I, Tsilimigkas G, Koukoulas S, Balatsos T (2014) The shift to competitiveness and a new phase of sprawl in the Mediterranean city: enterprises guiding growth in Messoghia—Athens. Cities 39:133–143
119. Salvati L, Ferrara C (2013) Do changes in vegetation quality precede urban sprawl? Area 45(3):365–375
120. Schewenius M, McPhearson T, Elmqvist T (2014) Opportunities for increasing resilience and sustainability of urban social-ecological systems: insights from the URBES and the cities and biodiversity outlook projects. Ambio 43(4):434–444
121. Gill SE, Handley JF, Roland Ennos A, Pauleit S, Theuray N, Lindley SJ (2008) Characterising the urban environment of UK cities and towns: a template for landscape planning. Landsc Urban Plan 87:210–222
122. Souliotis N (2013) Cultural economy, sovereign debt crisis and the importance of local contexts: the case of Athens. Cities 33:61–68
123. Karadimitriou N, Maloutas T, Arapoglou VP (2021) Multiple deprivation and urban development in Athens, Greece: spatial trends and the role of access to housing. Land 10(3):290
124. Grout C, Cavailhès J, Détang-Dessendre C, Thomas A (2016) Is sprawling residential behavior influenced by climate? Land Econ 92(2):203–219
125. Jomaa I, Auda Y, Abi Saleh B, Hamzé M, Safi S (2008) Landscape spatial dynamics over 38 years under natural and anthropogenic pressures in Mount Lebanon. Landsc Urban Plan 87:67–75
126. Kallis G (2010) Co-evolution in water resource development. The vicious cycle of water supply and demand in Athens, Greece. Ecol Econ 69(4):796–809
127. Rontos K, Grigoriadis E, Sateriano A, Syrmali M, Vavouras I, Salvati L (2016) Lost in protest, found in segregation: divided cities in the light of the 2015 "Οχι" referendum in Greece. City Cult Soc 7(3):139–148
128. Raudsepp-Hearne C, Peterson GD, Bennett EM (2010) Ecosystem service bundles for analyzing tradeoffs in diverse landscapes. Proc Natl Acad Sci USA 107(11):5242–5247
129. Radeloff VC, Hammer RB, Stewart SI, Fried JS, Holcomb SS, Mckeefry JF (2005) The wildland-urban interface in the United States. Ecol Appl 15(3):799–805
130. Gkartzios M, Remoundou K, Garrod G (2017) Emerging geographies of mobility: the role of regional towns in Greece's 'counterurbanisation story.' J Rural Stud 55:22–32
131. Ferrara A, Salvati L, Sabbi A, Colantoni A (2014) Urbanization, soil quality and rural areas: towards a spatial mismatch? Sci Total Environ 478:116–122
132. De Zeeuw H, Van Veenhuizen R, Dubbeling M (2011) The role of urban agriculture in building resilient cities in developing countries. J Agric Sci 149(S1):153–163
133. Economidou E (1993) The Attic landscape throughout the centuries and its human degradation. Landsc Urban Plan 24:33–37
134. Song Y, Knaap GJ (2004) Measuring urban form: is Portland winning the war on sprawl? J Am Plann Assoc 70(2):210–225
135. Duvernoy I, Zambon I, Sateriano A, Salvati L (2018) Pictures from the other side of the fringe: urban growth and peri-urban agriculture in a post-industrial City (Toulouse, France). J Rural Stud 57:25–35
136. Catalàn B, Sauri D, Serra P (2008) Urban sprawl in the Mediterranean? Patterns of growth and change in the Barcelona metropolitan region 1993–2000. Landsc Urban Plan 85(3–4):174–184
137. Colantoni A, Grigoriadis E, Sateriano A, Venanzoni G, Salvati L (2016) Cities as selective land predators? a lesson on urban growth, deregulated planning and sprawl containment. Sci Total Environ 545:329–339
138. Muñoz F (2003) Lock living: urban sprawl in Mediterranean cities. Cities 20:381–385
139. Paul V, Haslam Mckenzie F (2013) Peri-urban farmland conservation and development of alternative food networks: insights from a case study area in metropolitan Barcelona (Catalonia, Spain). Land Use Policy 30:94–105

140. Salvati L, Morelli VG, Rontos K, Sabbi A (2013) Latent exurban development: city expansion along the rural-to-urban gradient in growing and declining regions of southern Europe. Urban Geogr 34(3):376–394
141. Smiraglia D, Ceccarelli T, Bajocco S, Perini L, Salvati L (2015) Unraveling landscape complexity: land use/land cover changes and landscape pattern dynamics (1954–2008) in contrasting peri-urban and agro-forest regions of northern Italy. Environ Manage 56:916–932

Chapter 4
Local Development, Social Relations, and the 'Informal' Consolidation of Urban Voids

Abstract Focusing on social dynamics, spatial planning and governance deregulation in global urban systems, latent trends in working-classes over time and space were considered a powerful engine of metropolitan development. These reflections advocate a cautious generalization of urban specificities, emphasizing mechanism-based explanations transferable across various economic contexts. The resulting argument foregrounds interdependencies among regulation, market practices and everyday tactics that co-produce settlement forms, public spaces, and urban design. This contribution advances an interpretive claim regarding the socio-spatial dynamics under examination, situating the discussion within a broader (comparative) frame ultimately dealing with the formation (or consolidation) of urban voids.

Keywords Socio-demographic dynamics · Environmental quality · Vacant land · Settlement morphology · Planning

4.1 Introduction

Spatial planning and metropolitan governance are increasingly required to design new and original spaces – geographically distant from the past – with the hope to stay in tune with the surrounding landscapes [1–3]. Urban systems rarely drew up the meaning of schedule differences in functional macro-scale or the role of zoning processes [4]. In such contexts, new settlements are basically understood both as a form of physical space and evolving economic (and social) opportunities [5–7].

Conceptually, an empirical analysis of public space and urban design should propose a layered reading of causality in which institutional legacies, path-dependent preferences, and infrastructural endowments co-produce spatial outcomes [8–10]. This stance implies that spatial regularities should be evaluated as transient equilibriums rather than immutable laws underlying morphological complexity of urban landscapes [11–13]. The resulting argument foregrounds interdependencies among regulation, market practices and everyday tactics that produce (or co-produce) settlement forms [14–16].

L. Salvati and I. Konaxis, *Vacant Land*,
SpringerBriefs in Architectural Design and Technology,
https://doi.org/10.1007/978-981-95-4204-8_4

A form of elitist interpretation, reading intense dynamics of simplification, stadial evolution, and convergence to shapes, structures, and functions, has often permeated urban design [17]. At the same time, 'building spontaneity' has intended a key notion to read landscapes as economic space in regional contexts inclined to *laissez faire* and chaotic planning practices [18]. The consequent production of territories—partially out of planning control—has a profound impact on the structuring of metropolitan systems [19–22]. In this perspective, the present contribution advances an interpretive claim regarding socio-spatial dynamics and fringe land management (Fig. 4.1), situating the discussion within a broader comparative frame.

Fig. 4.1 Traditional rural settlements and fringe land. *Source* authors' photographic archive

4.2 Culture or Economy: The Basic Ground of Cities' Development

Informality is approached here not as a residual pathology, but as a patterned response to regulatory incongruence and credit scarcity, frequently operating through incremental construction and kinship-based finance [23–26]. Such practices instantiate a vernacular planning capacity that, while often extra-legal, negotiates access to centrality, services and political recognition (e.g. [27–29]). For instance, informal economy in the Mediterranean has reflected, on the one side, the traditional role of capitalistic markets, and the refuse of the state as the regulator of market dynamics, on the other side [30–32]). The picture is further consolidated by the lack of mechanisms which are capable of modifying correctly salaries. In this perspective, economic spontaneity and 'auto-promotion' are typical characteristics of Southern European societies [33–35].

Revisiting human ecology through a contemporary lens necessitates coupling ecological metaphors with asset-specific frictions, particularly those arising from land titling, mortgage markets and welfare regimes [36]. Accordingly, ecological processes such as 'succession' or 'invasion'—possibly regarded as appropriate modeling approaches to urban dynamics and metropolitan complexity—should be understood as mediated by policy design and the temporalities of infrastructure provision [37]. A correct interpretation of such dynamics, however, requires a coherent and comprehensive knowledge of socio-spatial patterns and economic characteristics of the places under examination [38–40].

Material infrastructures are read as socio-technical ensembles whose affordances reconfigure feasible urban forms and everyday mobility (e.g. [41]). Cultural repertoires, in turn, stabilize or destabilize these ensembles by legitimizing particular land uses and aesthetic preferences [42–44], in line with a flexible use of land – usually associated with the temporary exploitation of vacant land along fringe districts. In addition, climatic conditions seem to play a non-negligible role in shaping urban landscapes and metropolitan life-styles [45–49].

4.3 The Long-Lasting Linkage Between Urban Ideology and Fringe Land

Pro-urban disposition—especially in the European and Mediterranean regions, as well as in many other areas of the world—is interpreted as a pragmatic orientation toward resource pooling and opportunity density rather than as a mere preference for 'urban symbolism' [50]. Authoritarian interludes re-scripted such orientations by mobilizing 'ruralist' imaginaries, yet everyday practices often circumvented formal strictures through tactical urbanism [32, 51, 52]. The co-evolution of landscapes and social structures foregrounds the role of knowledge practices—agronomic, cadastral,

Fig. 4.2 The soil–water cropland interface in Mediterranean fringe land. *Source* authors' photographic archive

statistical—in translating environmental constraints into allocative decisions [53–55]. For instance, water–soil synergies function as critical thresholds (Fig. 4.2): once sealing and compaction exceed context-specific bounds, restoration costs rise non-linearly [56].

Monitoring frameworks should therefore privilege leading indicators—e.g. infiltration proxies and parcel-scale heterogeneity—over lagging measures of degradation [57–59]. Recognizing this co-evolution in vacant fringe land invites mixed-method strategies capable of integrating narrative evidence with geo-referenced indicators (e.g. [60–63]. Peri-urban districts with consolidated voids display recursive feedbacks whereby speculative expectations alter farm investment horizons, which in turn reshape fragmentation patterns and ecological connectivity [47, 64, 65].

4.4 Concluding Remarks

The proposed perspective reframes soil management, sprawl control, and optimal use of vacant land as a distributive question about who bears the costs of ecological restoration and infrastructural under-provision. Future research could formalize the mechanisms outlined here through comparative, longitudinal designs attentive to institutional reforms and credit cycles. An adaptive governance approach would sequence conservation, tenure consolidation and selective densification to reduce ecological edge effects.

References

1. Sassen S (2010) The city: its return as a lens for social theory. City Cult Soc 1(1):3–11
2. Swyngedouw E (2004) Social power and the urbanization of water. Oxford University Press, Oxford, UK
3. Zambon I, Colantoni A, Salvati L (2019) Horizontal vs vertical growth: understanding latent patterns of urban expansion in large metropolitan regions. Sci Total Environ 654:778–785
4. Gonzales S (2011) The rth/south divide in Italy and England: discursive construction of regional inequality. Eur Urban RegNal Stud 18(1):62–76
5. Di Feliciantonio C, Salvati L (2015) 'Southern' alternatives of urban diffusion: investigating settlement characteristics and socio-economic patterns in three Mediterranean regions. Tijdschr Econ Soc Geogr 106(4):453–470
6. Liu L, Meng L (2020) Patterns of urban sprawl from a global perspective. J Urban Plan Dev 146(2):0402
7. Vidal M, Domene E, Saurí D (2011) Changing geographies of water–related consumption: residential swimming pools in suburban Barcelona. Area 43(1):67–75
8. Cassiers T, Kesteloot C (2012) Socio–spatial inequalities and social cohesion in European cities. Urban Studies 49(9):1909–1924
9. Gospodini A (2009) Post–industrial trajectories of Mediterranean European cities: the case of post–Olympics Athens. Urban Stud 46(5–6):1157–1186
10. Townsend A (2009) Integration of economic and spatial planning across scales. Int J Public Sect Manag 22(7):643–659
11. Domene E, Sauri D, Parés M (2005) Urbanization and sustainable resource use: the case of garden watering in the metropolitan region of Barcelona. Urban Geogr 26(6):520–535
12. Karadimitriou N, Maloutas T, Arapoglou VP (2021) Multiple deprivation and urban development in Athens, Greece: spatial trends and the role of access to housing. Land 10(3):290
13. Koprowska K, Łaszkiewicz E, Kronenberg J (2020) Is urban sprawl linked to green space availability? Ecol Ind 108(10):5723
14. Chorianopoulos I, Pagonis T, Koukoulas S, Drymoniti S (2010) Planning, competitiveness and sprawl in the Mediterranean city: the case of Athens. Cities 27:249–259
15. Garcia-López MA, Muñiz I (2010) Employment decentralisation: polycentricity or scatteration? Case Barcelona Urban Stud 47(14):3035–3056
16. Lee S (2011) Metropolitan growth patterns and socio–economic disparity in six US Metropolitan areas 1970–(2000). Int J Urban RegNal Res 35(5):988–1011
17. Mouratidis K (2019) Compact city, urban sprawl, and subjective well–being. Cities 92:261–272
18. Lamonica GR, Recchioni MC, Chelli FM, Salvati L (2020) The efficiency of the cross–entropy method when estimating the technical coefficients of input–output tables. Spat Econ Anal 15:62–91
19. Orenstein DE, Frenkel A, Jahshan F (2014) Methodology matters: measuring urban spatial development using alternative methods. Environ Plann B Plann Des 41:3–23
20. Salvati L (2014) Towards a polycentric region? the socioeconomic trajectory of Rome, an 'eternal' Mediterranean City. Tijdschr Econ Soc Geogr 105(3):268–284
21. Salvati L, Ferrara A, Chelli F (2018) Long–term growth and metropolitan spatial structures: An analysis of factors influencing urban patch size under different economic cycles. Geogr Tidsskr – Dan J Geogr 118(1):56–71
22. Tsilimigkas G, Kizos T (2014) Space, pressures and the management of the Greek landscape. Geogr Ann: Ser B, Hum Geogr 96(2):159–175

23. Allen J, Barlow J, Maloutas T, Padovan L (2004) Housing in Southern Europe. Blackwell, London
24. Masini E, Tomao A, Barbati A, Corona P, Serra P, Salvati L (2019) Urban growth, land–use efficiency and local socioeconomic context: a comparative analysis of 417 metropolitan regions in Europe. Environ Manage 63:322–337
25. Scott AJ, Storper M (2015) The nature of cities: the scope and limits of urban theory. Int J Urban Reg Res 39(1):1–15
26. Storper M, Scott AJ (2016) Current debate in urban theory: a critical assessment. Urban Studies 53(6):1114–1136
27. Andreotti A, Le Galès P, Moreno Fuentes FJ (2013) Controlling the urban fabric: the complex game of distance and proximity in European upper–middle–class residential strategies. Int J Urban Reg Res 37(2):576–597
28. Durà-Guimera A (2003) Population deconcentration and social restructuring in Barcelona, a European Mediterranean city. Cities 20(6):387–394
29. Musterd S (2005) Social and Ethnic Segregation in Europe: levels, Causes and Effects. J Urban Aff 27(3):331–348
30. Arbaci S, Malheiros J (2010) De–segregation, Peripheralisation and the social exclusion of immigrants: southern European cities in the 1990s. J Ethn Migr Stud 36(2):227–255
31. Maloutas T (2004). Segregation and residential mobility. Socially entrapped social mobility and its impact on segregation in Athens. Eur Urban RegNal Stud 11(2):171–187
32. Salvati L, Morelli VG, Rontos K, Sabbi A (2013) Latent exurban development: city expansion along the rural–to–urban gradient in growing and declining regions of southern Europe. Urban Geogr 34(3):376–394
33. Buzar S, Ogden P, Hall R, Haase A, Kabisch S, Steinfiihrer A (2007) Splintering urban populations: emergent landscapes of Reurbanisation in four European cities. Urban Studies 44(4):651–677
34. Couch C, Petschel-Held G, Leontidou L (2007) Urban sprawl in Europe: landscapes, land–use change and policy. Blackwell, Oxford
35. Zambon I, Benedetti A, Ferrara C, Salvati L (2018) Soil matters? a multivariate analysis of socioeconomic constraints to urban expansion in Mediterranean Europe. Ecol Econ 146:173–183
36. Galster G, Hanson R, Ratcliffe MR, Wolman H, Coleman S, Freihage J (2001) Wrestling sprawl to the ground: defining and measuring an elusive concept. Hous Policy Debate 12(4):681–717
37. Domene E, Sauri D (2006) Urbanization and water consumption: influential factors in the metropolitan region of Barcelona. Urban Studies 43(9):1605–1623
38. Arapoglou VP, Sayas J (2009) New facets of urban segregation in Southern Europe. Eur Urban RegNal Stud 16(4):345–362
39. Arbaci S (2008) (Re) viewing ethnic residential segregation in southern European cities: housing and urban regimes as mechanisms of marginalisation. Hous Stud 23(4):589–613
40. Ciommi M, Chelli FM, Salvati L (2019) Integrating parametric and n–parametric multivariate analysis of urban growth and commuting patterns in a European metropolitan area. Qual Quant 53(2):957–979
41. Richardson H, Chang-Hee C (2004) Urban sprawl in Western Europe and the United States. Ashgate, London
42. Carlucci M, Chelli FM, Salvati L (2018) Toward a new cycle: Short–term population dynamics, gentrification, and re–urbanization of Milan (Italy). Sustainability 10(9):3014
43. Keddie J, Tonkiss F (2010) The market and the plan: housing, urban renewal and socio–economic change in London. City Cult Soc 1(2):57–67
44. Musterd S, Marcińczak S, Van Ham M, Tammaru T (2017). Socioeconomic segregation in European capital cities. Increasing separation between poor and rich. Urban Geography 38(7):1062–1083
45. Angel S, Parent J, Civco DL, Blei A, Potere, D (2011). The dimensions of global urban expansion: estimates and projections for all countries: 2000–2050 Progress in Planning 75(2):53–107

46. Salvati L, Zitti M (2009) Substitutability and weighting of ecological and economic indicators: exploring the importance of various components of a synthetic index. Ecol Econ 68–4:1093–1099
47. Salvati L, Petitta M, Ceccarelli T, Perini L, Di Battista F, Scarascia MEV (2008) Italy's renewable water resources as estimated on the basis of the monthly water balance. Irrig Drain: J Int Comm Irrig Drain 57(5):507–515
48. Vinci S, Egidi G, López Gay A, Salvati L (2022) Population growth and urban management in metropolitan regions: the contribution of natural balance and migration to polycentric development in Barcelona. Appl Spat Anal Policy 15(1):71–94
49. Vinci S, Vardopoulos I, Salvati L (2023) A tale of a shrinking City? exploring the complex interplay of socio–demographic dynamics in the recent development of Attica Greece. Cities 132(10):4089
50. Salvati L, Carlucci M (2016) Patterns of sprawl: the socioeconomic and territorial profile of dispersed urban areas in Italy. Reg Stud 50(8):1346–1359
51. Bueno-Suárez C, Coq-Huelva D (2020) Sustaining what is unsustainable: a review of urban sprawl and urban socio–environmental policies in rth America and Western Europe. Sustainability 12(11):4445
52. Leeuwen T (2000) The Springboard in the Pond: an intimate history of the swimming pool. Graham Foundation/MIT Press Series in Contemporary Architectural Discourse, Boston
53. Cecchini M, Zambon I, Pontrandolfi A, Turco R, Colantoni A, Mavrakis A, Salvati L (2019) Urban sprawl and the 'olive' landscape: sustainable land management for 'crisis' cities. GeoJournal 84(1):237–255
54. Duvernoy I, Zambon I, Sateriano A, Salvati L (2018) Pictures from the other side of the fringe: Urban growth and peri–urban agriculture in a post–industrial city (Toulouse, France). J Rural Stud 57:25–35
55. Nickayin SS, Tomao A, Quaranta G, Salvati L, Gimenez Morera A (2020) Going toward resilience? town planning, peri-urban landscapes, and the expansion of Athens Greece. Sustainability 12(24):0471
56. Bruegmann R (2019) Sprawl: a compact history. University of Chicago Press, Chicago
57. Chelleri L, Schuetze T, Salvati L (2015) Integrating resilience with urban sustainability in neglected neighborhoods: challenges and opportunities of transitioning to decentralized water management in Mexico City. Habitat Int 48:122–130
58. Salvati L, Ridolfi E, Pujol DS, Ruiz PS (2016) Latent sprawl, divided Mediterranean landscapes: Urban growth, swimming pools, and the socio–spatial structure of Athens Greece. Urban Geogr 37(2):296–312
59. Zambon I, Colantoni A, Carlucci M, Morrow N, Sateriano A, Salvati L (2017) Land quality, sustainable development and environmental degradation in agricultural districts: a computational approach based on entropy indexes. Environ Impact Assess Rev 64:37–46
60. Bajocco S, Ceccarelli T, Smiraglia D, Salvati L, Ricotta C (2016) Modeling the ecological niche of long–term land use changes: the role of biophysical factors. Ecol Ind 60:231–236
61. Kosmas C, Karamesouti M, Kounalaki K, Detsis V, Vassiliou P, Salvati L (2016) Land degradation and long–term changes in agro–pastoral systems: an empirical analysis of ecological resilience in Asteroussia-Crete (Greece). CATENA 147:196–204
62. Kroll F, Kabisch N (2012) The relation of diverging urban growth processes and demographic change along an urban-rural gradient. Popul Space Place 18:260–276
63. Smiraglia D, Ceccarelli T, Bajocco S, Perini L, Salvati L (2015) Unraveling landscape complexity: land use/land cover changes and landscape pattern dynamics (1954–2008) in contrasting peri-urban and agro-forest regions of rthern Italy. Environ Manage 56(4):916–932
64. Biasi R, Brunori E, Smiraglia D, Salvati L (2015) Linking traditional tree–crop landscapes and agro–biodiversity in Central Italy using a database of typical and traditional products: a multiple risk assessment through a data mining analysis. Biodivers Conserv 24(12):3009–3031
65. Zwiers M, Bolt G, van Ham M, Van Kempen R (2016) The global financial crisis and neighbourhood decline. Urban Geogr 37(5):664–684

66. Alexandri G (2015) Reading between the lines: gentrification tendencies and issues of urban fear in the midst of Athens' crisis. Urban Studies 52(9):1631–1646
67. Garcia-López MÀ (2019) All roads lead to Rome... and to sprawl? evidence from European cities. Reg Sci Urban Econ 79:10–3467
68. Pacione M (2003) Urban environmental quality and human wellbeing—a social geographical perspective. Landsc Urban Plan 65:19–30
69. Parés M, March H, Saurí D (2013) Atlantic gardens in Mediterranean climates: understanding the production of suburban natures in Barcelona. Int J Urban Reg Res 37:328–347
70. Perrin C, Nougarèdes B, Sini L, Branduini P, Salvati L (2018) Governance changes in peri–urban farmland protection following decentralisation: a comparison between Montpellier (France) and Rome (Italy). Land Use Policy 70:535–546
71. Salvati L, Ciommi MT, Serra P, Chelli FM (2019) Exploring the spatial structure of housing prices under economic expansion and stagnation: the role of socio–demographic factors in metropolitan Rome, Italy. Land Use Policy 81:143–152
72. Salvati L, Zitti M (2007) Territorial disparities, natural resource distribution, and land degradation: a case study in Southern Europe. GeoJournal 70–2:185–194
73. Salvati L, Gemmiti R, Perini L (2012) Land degradation in Mediterranean urban areas: an unexplored link with planning? Area 44(3):317–325
74. Salvati L, Zitti M, Sateriano A (2013) Changes in city vertical profile as an indicator of sprawl: evidence from a Mediterranean urban region. Habitat Int 38:119–125
75. Zambon I, Serra P, Sauri D, Carlucci M, Salvati L (2017) Beyond the 'Mediterranean city': socioeconomic disparities and urban sprawl in three Southern European cities. Geogr Ann: Ser B, Hum Geogr 99(3):319–337

Chapter 5
Fallow Land and 'Vacant' Peri-Urban Spaces

Abstract Urban expansion, economic growth, and transportation infrastructure exert significant pressures on landscapes, particularly in suburban regions, leading to farmland abandonment and the proliferation of fallow land. These areas, intended as agricultural land left unsown to restore fertility or due to shifting production incentives, represent both an ecological resource and a latent asset within urbanizing territories. Their spatial distribution reflects complex interactions between economic cycles, land-use policies, and speculative pressures of real estate market, highlighting the dual role of fallow land as a 'buffer zone' for agricultural activity and as a stock for urban development. Sequential economic downturns influence the dynamics of fallow land, with expansion periods often promoting land withholding for speculation, while downturns enable re-cultivation and intensification of traditional crops. These processes are further shaped by planning regulations, zoning flexibility, and informal land-use practices, contributing to spatial heterogeneity in metropolitan expansion and thus consolidating the entropic distribution of vacant land. Fallow land also plays a critical ecological role, supporting soil conservation, water regulation, biodiversity, and landscape connectivity, while maintaining cultural and agricultural heritage. Effective management of peri-urban landscapes requires integrated, multilevel strategies that consider socioeconomic, environmental, and cultural dimensions together. Recognizing fallow land as a practical example of peri-urban vacant land and, at the same time, as a strategic component of urban–rural interfaces, allows for sustainable land-use planning, balancing agricultural productivity, ecological resilience, and urban containment. This approach provides a framework for understanding and guiding landscape transformations in advanced economies experiencing rapid urbanization, social transformations, and land-use change.

Keywords Urban–rural fringe · Local development · Land management · Agriculture · Sustainability

L. Salvati and I. Konaxis, *Vacant Land*,
SpringerBriefs in Architectural Design and Technology,
https://doi.org/10.1007/978-981-95-4204-8_5

5.1 Introduction

The progressive transformation of agricultural landscapes through settlement expansion constitutes one of the most profound and enduring ecological pressures on contemporary territories (e.g. [1]). Nowhere is this process more visible, nor its implications more acute, than in metropolitan and peri-urban regions, where the logic of urbanization directly competes with the preservation of agricultural land and ecosystem resources [2]. Policy frameworks addressing farmland consumption have traditionally relied upon generalized prescriptions and broad-scale regulatory mechanisms (e.g. [3–5]). However, the empirical evidence of earlier studies suggests that the efficacy of such strategies is contingent upon their spatial precision. The reduction of land take should be thus oriented toward those territories where ecological functions are really vulnerable, and/or where land-use transitions generate disproportionate socio-environmental costs leading to cropland abandonment and the economic erosion of agricultural resources [6]. This requires systematic knowledge regarding the consequences of specific land conversions that improve soil performance, hydrological equilibrium, and biodiversity integrity [7].

Under contemporary history, farmland abandonment epitomizes one of the most distinctive land-use transitions in advanced economies. Driven by rising yields in high-productivity areas, the implementation of conservation-oriented measures, and an increasing reliance on global markets for agricultural imports, farmland abandonment reflects the multi-scalar reconfiguration of agrarian systems [8–10]. Yet its ecological and productive implications remain only partially understood. A key limitation stems from the absence of comprehensive data on the spatial distribution of fallow and abandoned land, particularly across large geographical extents (e.g. [11]). Building a comparative framework that explicitly distinguishes fallow land, abandoned farmland, and re-cultivated fields, is indispensable for assessing the ecological trajectories of abandonment and re-use, especially in peri-urban districts [12].

5.2 Fallow Land and Peri-Urban Farming

The category of fallow land—long embedded within agricultural practice and theory—provides an analytical lens through which land-use intensity and management options can be investigated. By definition, fallow denotes land that has been tilled yet deliberately left unsown to regenerate fertility or avert overproduction. While being intended as an indirect example of vacant land in peri-urban districts, its extension operates as an implicit indicator of regional land-use intensity and agricultural adjustment [13]. Being commonly associated with extensive systems on marginal soils, fallow land emerged with striking frequency in fringe settlements, where agricultural abandonment is driven less by ecological limits than by real-estate speculation and expectations of future (urban) development [14]. In these cases, the very act of leaving land uncultivated is embedded in a broader political economy of land valuation and metropolitan expansion [15].

Within the European Union, agri-environmental schemes under the Common Agricultural Policy (CAP) have institutionalized fallow land as a critical policy variable, framing it simultaneously as a tool for ecological stewardship and as an indicator of rural development [16–18]. The share of fallow land in total landscape functions as a diagnostic metric, one that captures processes of agricultural extensivation, while distinguishing between the immediate effects of agronomic practices ('performances') and the broader environmental consequences they entail ('outcomes'). A longitudinal analysis of fallow dynamics at local and regional scales further illuminates the persistence (or decline) of traditional management practices [19–21], thereby situating land-use change within historically contingent trajectories of rural adaptation (Fig. 5.1).

Realizing a truly sustainable land management is particularly needed in metropolitan contexts, where landscape transformations are both rapid and complex, and reflect the latent interaction of economic restructuring, demographic pressure, and institutional frameworks [22–25]. Land-use change in such settings should not be reduced to deterministic trends, rather, it constitutes the outcome of contested social processes involving diverse actors and competing rationalities [26]. In this perspective, European cities since the early 1970s have undergone a paradigmatic shift

Fig. 5.1 Mixed landscapes at the fringe: residential sprawl (left) and fallow land (right). *Source* authors' photographic archive

Fig. 5.2 The spatial conflict between fallow land and fringe infrastructure. *Source* authors' photographic archive

from compact, dense morphologies to increasingly fragmented settlement structures characterized by peri-urban sprawl [27–29]. The trajectory from 'compact growth' to 'spatially discontinuous urbanization' has reconfigured peri-urban agricultural systems in uneven and spatially differentiated ways [30, 31], reshaping the incidence of fallow land and altering the viability of specific crops (e.g. [32]).

5.2.1 Fallow Land and Economic Downturns

Earlier studies have demonstrated how fallow land in peri-urban districts can be conceptualized as a dual resource. On the one hand, it represents a latent economic stock, susceptible to speculative dynamics and potentially subject to reclassification through zoning modifications that would enable urban development [33]. On the other hand, it constitutes a contingent stock of cultivable land, which can be mobilized during phases of economic stagnation when construction activity contracts and agricultural intensification becomes relatively more attractive (e.g. [34]). By situating fallow land at the intersection of speculation and production, this interpretation makes the latent interplay between cycles of urban expansion and agrarian adjustment fully explicit [35]. More broadly, it provides an interpretative framework for understanding landscape transformations in metropolitan contexts [36], where comparable ecological constraints and territorial dynamics are at play.

Within the conjuncture of economic crisis, a small yet expanding cohort of urban residents publicly articulated their willingness to exit the urban labor market in order to seek livelihoods in the agricultural sector, signaling a partial revaluation of farming as a crisis-adaptive strategy [37]. At the same time, earlier inquiries into multi-activity within peri-urban households revealed that nearly half of working-age individuals engaged in multiple occupations (e.g. [38]). This condition reflected the relative abundance of alternative employment niches available in fringe districts when compared to rural hinterlands [39], underscoring the inherent complexity in peri-urban socioeconomic systems [40].

The cyclical expansion and contraction of fallow land could therefore be situated within the broader oscillations of macroeconomic growth and recession [41]. During periods of economic expansion, fallow land expanded and stabilized, against a more evident decline in total cropland (e.g. [42]). Such areas of fallow were predominantly located along the fringe, where agriculture was most severely confronted with urban encroachment and where land values were subject to the highest levels of speculative volatility [43]. By contrast, in phases of economic contraction, the retrenchment of land speculation coincided with a reduction of fallow areas in more accessible rural districts, which were reconverted into higher-value crops [44]. Fallow land thus functioned as a latent source of free, unconstrained soils whose exploitation alternated between speculative building (residential, industrial) purposes during booms, and productive agricultural uses during recessions (e.g. [45]).

5.3 Sustainable Soil Management, Spatial Planning and Vacant (Fallow) Land

In this light, the regulation of land-use dynamics in peri-urban areas has been increasingly conceptualized as a priority of soil protection and agricultural policy [46]. Yet empirical analyses across diverse socioeconomic conditions remain limited, leaving significant gaps in the comparative understanding of these processes [47]. Landscape transformations have historically unfolded with particular intensity in local contexts where suburbanization generates rapid demographic expansion and where ecological systems are inherently fragile due to specific geomorphological and land-use conditions [48]. These dynamics have resulted in the progressive consumption of soils and the destabilization of rural territories—processes that have, in turn, produced cumulative and persistent (negative) externalities on local environments (Fig. 5.3).

Fig. 5.3 Relict, aesthetically degraded agricultural landscapes mixing horticulture and fallow land at the fringe of coastal settlements. *Source* authors' photographic archive

Land speculation was reinforced by the structural weakness of planning institutions: the excessive flexibility of land-zoning regulations, the systematic ineffectiveness of metropolitan master plans, and recurrent cycles of unauthorized construction (e.g. later legalized in pre-electoral periods, as frequently observed in Mediterranean countries during the 1960s and the 1970s), all of which contributed to institutionalizing land-use volatility [49–51]. The global financial crisis of 2007 decisively disrupted this consolidated trajectory, curtailing construction activity, and recalibrating the interplay between speculative and productive land uses [52].

Parallel to these shifts, rural land-use trajectories revealed increasing spatial polarization. Empirical evidence documents a growing divide between marginal rural lands—where abandonment, forest encroachment, rising fallow shares, and diffuse sprawl coalesce (Fig. 5.4)—and highly productive agricultural districts, which maintained stable cropland, minimized fallow extension, and exhibited lower susceptibility to speculative conversion [53–55]. Such patterns mirror broader landscape restructuring processes, possibly including localized agricultural intensification (e.g. expansion of vineyards and horticultural crops at the expense of arable land) and extensivation (e.g. expansion of olive groves in Mediterranean regions and rain-fed cereals in Northern countries).

For instance, olive groves have been specifically recognized for their capacity to buffer human pressure, ensuring ecological stability through soil conservation and

Fig. 5.4 Relict agro-forest systems under progressive abandonment. *Source* authors' photographic archive

the preservation of water resources [56]. Nonetheless, the continuing transformation of peri-urban landscapes into low-quality agricultural or pasture land remains a structural driver of ecological fragility, undermining the resilience of fertile soils and compromising agro-ecosystem connectivity [57–60].

5.3.1 Fallow Land and Social Dynamics

In this light, fallow land assumes analytical significance as a sensitive proxy of socio-spatial restructuring. The erosion of full-time farming, the decline of large-scale holdings, and the increasing prevalence of irrigation, mechanization, and fragmented labor practices all intersect with macroeconomic cycles to reshape peri-urban agro-ecosystems [61]. Economic instability in turn accentuated structural divergences with heavily industrialized peri-urban environments, albeit under increasing tension between profitability imperatives and rural identity [62]. Grounded on such reasons, landscape evolution resulted in hybrid socio-ecological configurations—where urban and rural logics coexist in unstable equilibriums [63–65].

Within this co-evolutionary dynamic, fallow land emerges as the most responsive element, simultaneously representing a speculative reserve for future development and an immediately mobilizable stock for agricultural intensification (Fig. 5.5). From a governance perspective, this dualism underscores the urgency of designing multi-scalar land-use planning strategies that stabilize fallow land [66]. Current subsidy regimes only partially address this imperative: the spatial distribution of subsidized fallow diverges substantially from the broader distribution of abandoned land, thereby stabilizing land-use trajectories in some contexts while inadvertently reinforcing speculative pressures and accelerating conversion to urban uses in others [67–69].

Given the simultaneous advance of forest and shrubland cover and the gradual erosion of farmland in advanced economies, the deliberate promotion of agroforestry systems—characterized by multifunctional land use that integrates tourism, wildlife conservation, hunting, recreational activities, extensive livestock grazing, and timber production—represents an indirect yet highly effective means of sustaining ecological potential in peri-urban districts [70–72]. At the same time, such practices contribute to the preservation of cultural traditions historically intertwined with landscapes in which fallow land has long played a central role (e.g. [73]).

Fig. 5.5 Relict woodlots progressively abandoned after repeated burning. *Source* authors' photographic archive

5.4 Concluding Remarks

Regional science benefits from a trans-disciplinary vision that combines architectural disciplines with geographical and ecological aspects of landscapes and settlements. In this vein, our work tries to represent the complex geography of urban–rural relationships, drawing original ideas for analysis, reflection and design. Conceived as both a reservoir and a buffer of agriculturally productive soils throughout alternating phases of economic growth and decline, the spatio-temporal dynamics of fallow land constitute a sensitive indicator of the resilience and adaptive capacity of peri-urban districts. Assuming the decline in ecosystem services as a symptom of mismanagement practices, the safeguarding of fallow land against ecological decline emerges as a critical priority for rural development strategies.

The inherent territorial heterogeneity, coupled with the topographic complexity of metropolitan settings and the accelerating fragmentation of land induced by urbanization, pose additional, significant challenges for political authorities, policymakers, and stakeholders interested in vacant soil management. If one accepts the premise that regional transformations exert feedbacks upon local socioeconomic systems, then the resulting environmental configurations necessitate systematic monitoring and the articulation of policy responses grounded in socio-ecological reasoning. In this perspective, the governance of landscape transformations cannot be adequately achieved through short-term interventions alone, but requires the design and implementation of multi-scalar, integrated strategies. Such strategies should simultaneously account for cyclical economic fluctuations (periods of expansion and recession) and for long-term structural processes, thereby aligning immediate socioeconomic priorities with broader, enduring trajectories of territorial reorganization.

References

1. Agnoletti M, Cargnello G, Gardin L, Santoro A, Bazzoffi P, Sansone L, Pezza L, Belfiore N (2011) Traditional landscape and rural development: Comparative study in three terraced areas in northern, central and southern Italy to evaluate the efficacy of GAEC standard 4.4 of cross compliance. Ital J Agron 6:121–139
2. Biasi R, Brunori E, Smiraglia D, Salvati L (2015) Linking traditional tree-crop landscapes and agro-biodiversity in Central Italy using a database of typical and traditional products: a multiple risk assessment through a data mining analysis. Biodivers Conserv 24(12):3009–3031
3. Amato F, Maimone BA, Martellozzo F, Nolè G, Murgante B (2016) The effects of urban policies on the development of Urban areas. Sustainability 8:297
4. Aubry RJ, Abat MH, Ramamonjisoa J, Dabat MH, Rakotoarisoa J, Rakotondraibe J, Rabeharisoa L (2012) Urban agriculture and land use in cities: an approach with multi-functionality and sustainability concepts in the case of Antananarivo (Madagascar). Land Use Policy 29:429–439
5. Brunori G, Rossi A (2007) Differentiating countryside. social representations and governance patterns in rural areas with high social density: the case of Chianti Italy. J Rural Stud 23:183–205
6. Tsilimigkas G, Kizos T (2014) Space, pressures and the management of the Greek landscape. Geogr Ann: Ser B, Hum Geogr 96(2):159–175

7. Nuissl H, Haase D, Lanzendorf M, Wittmer H (2009) Environmental impact assessment of urban land use transitions—a context-sensitive approach. Land Use Policy 26(2):414–424
8. Portnov BA, Safriel UN (2004) Combating desertification in the Negev: Dryland agriculture vs. dryland urbanization. J Arid Environ 56:659–680
9. Radeloff VC, Hammer RB, Stewart SI, Fried JS, Holcomb SS, Mckeefry JF (2005) The wildland-urban interface in the United States. Ecol Appl 15(3):799–805
10. Savo V, Caneva G, McClatchey W, Reedy D, Salvati L (2014) Combining environmental factors and agriculturalists' observations of environmental changes in the traditional terrace system of the Amalfi Coast (Southern Italy). Ambio 43(3):297–310
11. Salvati L, Ferrara A (2014) Do land cover changes shape sensitivity to forest fires in peri-urban areas? Urban For Urban Green 13(3):571–575
12. Estel S, Kuemmerle T, Alcántara C, Levers C, Prishchepov A, Hostert P (2015) Mapping farmland abandonment and recultivation across Europe using MODIS NDVI time series. Remote Sens Environ 163:312–325
13. Kuemmerle T, Erb K, Meyfroidt P, Müller D, Verburg PH, Estel S, Levers C (2013) Challenges and opportunities in mapping land use intensity globally. Curr Opin Environ Sustain 5(5):484–493
14. Grekousis G, Manetos P, Photis YN (2013) Modeling urban evolution using neural networks, fuzzy logic and GIS: the case of the Athens metropolitan area. Cities 30:193–203
15. Kizos T (2010) Multifunctionality of farm households in Greece. Nor Geogr Tidsskrift—Norwegian J Geogr 64(2):105–116
16. Kizos T, Vasdeki M, Chatzikiriakou C, Dimitriou D (2011) 'For my children': Different functions of the agricultural landscape and attitudes of farmers on different areas of Greece towards small scale landscape change. Geogr Tidsskr-Dan J Geogr 111(2):117–130
17. Kristensen SP, Thenail C, Kristensen L (2001) Farmers' involvement in landscape: an analysis of the relationship between farm location, farm characteristics and landscape changes in two study areas in Jutlund, Denmark. J Environ Manage 61:301–318
18. Primdahl J, Peco B, Schramek J, Andersen E, Oñate JJ (2003) Environmental effects of agri-environmental schemes in Western Europe. J Environ Manage 67(2):129–138
19. Kelly C, Ferrara A, Wilson GA, Ripullone F, Nolè A, Harmer N, Salvati L (2015) Community resilience and land degradation in forest and shrubland socio-ecological systems: evidence in Gorgoglione, Basilicata, Italy. Land Use Policy 46:11–20
20. Kizos T, Vakoufaris H (2011) Valorisation of local asset: the case of olive oil on Lesvos Island. Greece. Food Policy 36(5):705–714
21. Kosmas C, Karamesouti M, Kounalaki K, Detsis V, Vassiliou P, Salvati L (2016) Land degradation and long-term changes in agro-pastoral systems: an empirical analysis of ecological resilience in Asteroussia - Crete (Greece). CATENA 147:196–204
22. Colantoni A, Ferrara C, Perini L, Salvati L (2015) Assessing trends in climate aridity and vulnerability to soil degradation in Italy. Ecol Ind 48:599–604
23. Colantoni A, Grigoriadis E, Sateriano A, Sarantakou E, Salvati L (2017) Back to Von Thunen: a Southern European perspective on mono-centric urban growth, economic structure and non-urban land decline. Int Plan Stud 22(3):173–188
24. Garcia Latorre J, Garcia-Latorre J, Sanchez-Picon A (2001) Dealing with aridity: socio-economic structures and environmental changes in an arid Mediterranean region. Land Use Policy 18:53–64
25. Tanrivermis H (2003) Agricultural land use change and sustainable use of land resources in the Mediterranean region of Turkey. J Arid Environ 54:553–564
26. Cakir G, Un C, Baskent EZ, Kose S, Sivrikaya F, Keles S (2008) Evaluating urbanization, fragmentation and land use/cover change pattern in Istanbul city, Turkey from 1971 to 2002. Land Degrad Dev 19:663–675
27. Beriatos E, Gospodini A (2004) Glocalising urban landscapes: Athens and the 2004 Olympics. Cities 21:187–202
28. Bhatta B (2010) Causes and consequences of urban growth and sprawl. Analysis of urban growth and sprawl from remote sensing data. Springer, Berlin

29. Schneider A, Woodcock CE (2008) Compact, dispersed, fragmented, extensive? a comparison of urban growth in twenty-five global cities using remotely sensed data, pattern metrics and census information. Urban Studies 45(3):659–692
30. Serra P, Pons X, Saurì D (2008) Land-cover and land-use change in a Mediterranean landscape: a spatial analysis of driving forces integrating biophysical and human factors. Appl Geogr 28(3):189–209
31. Serra P, Vera A, Tulla AF, Salvati L (2014) Beyond urban-rural dichotomy: exploring socioeconomic and land-use processes of change in Spain (1991–2011). Appl Geogr 55:71–81
32. Seto KC, Ramankutty N (2016) Hidden linkages between urbanization and food systems. Science 352(6288):943–945
33. Smiraglia D, Ceccarelli T, Bajocco S, Salvati L, Perini L (2016) Linking trajectories of land change, land degradation processes and ecosystem services. Environ Res 147:590–600
34. Souliotis N (2013) Cultural economy, sovereign debt crisis and the importance of local contexts: the case of Athens. Cities 33:61–68
35. Audsley E, Pearn KR, Simota C, Cojocaru G, Koutsidou E, Rounsevell MDA, Trnka M, Alexandrov V (2006) What can scenario modelling tell us about future European scale agricultural land use and what not? Environ Sci Policy 9(2):148–162
36. Aretano R, Petrosillo I, Zaccarelli N, Semeraro T, Zurlini G (2013) People perception of landscape change effects on ecosystem services in small Mediterranean islands: a combination of subjective and objective assessments. Landsc Urban Plan 112(1):63–73
37. Gkartzios M (2013) ‘Leaving Athens’: narratives of counterurbanisation in times of crisis. J Rural Stud 32:158–167
38. Moissidis A, Duquenne M-N (1997) Peri-urban rural areas in Greece: the case of Attica. Sociol Rural 37:228–239
39. Gkartzios M, Remoundou K, Garrod G (2017) Emerging geographies of mobility: the role of regional towns in Greece’s ‘counterurbanisation story.’ J Rural Stud 55:22–32
40. Gkartzios M, Scott K (2015) A cultural panic in the province? Counterurban Mobilities, creativity, and crisis in greece. Popul Space Place 21(8):843–855
41. Delladetsima PM (2006) The emerging property development pattern in Greece and its impact on spatial development. Eur Urban RegNal Stud 13(3):245–278
42. De Zeeuw H, Van Veenhuizen R, Dubbeling M (2011) The role of urban agriculture in building resilient cities in developing countries. J Agric Sci 149(S1):153–163
43. Gospodini A (2009) Post-industrial trajectories of Mediterranean European cities: the case of post-Olympics Athens. Urban Studies 46:1157–1186
44. Loumou A, Giourga C (2003) Olive groves: ‘The life and identity of Mediterranean.’ Agric Hum Values 20:87–95
45. Salvati L, Ferrara C (2013) Do changes in vegetation quality precede urban sprawl? Area 45(3):365–375
46. Salvati L (2014) Agro-forest landscape and the ‘fringe’city: a multivariate assessment of land-use changes in a sprawling region and implications for planning. Sci Total Environ 490:715–723
47. Orsini S (2013) Landscape polarisation, hobby farmers and valuable hills in Tuscany: understanding landscape dynamics in a peri-urban context. Geogr Tidsskr-Dan J Geogr 113:54–64
48. Palazzo AL, Aristone O (2017) Peri-Urban matters. Changing olive growing patterns in central Italy. Sustainability 9(4):638
49. Foley JA, Ramankutty N, Brauman KA, Cassidy ES, Gerber JS, Johnston M, Mueller ND, O’Connell C, Ray DK, West PC, Balzer C, Bennett EM, Carpenter SR, Hill J, Monfreda C, Polasky S, Rockstrom J, Sheehan J, Siebert S, Tilman D, Zaks DPM (2011) Solutions for a cultivated planet. Nature 478:337–342
50. Olsson EGA, Kerselaers E, Søderkvist Kristensen L, Primdahl J, Rogge E, Wästfelt A (2016) Peri-Urban food production and its relation to urban resilience. Sustainability 8(12):1340
51. Salvati L (2016) From sprawl to compactness and back: long-term population dynamics (1848–2011) and the economic structure of a Mediterranean city. GeoJournal 81(2):319–332

52. Salvati L, Zitti M, Sateriano A (2013) Changes in city vertical profile as an indicator of sprawl: evidence from a Mediterranean urban region. Habitat Int 38:119–125
53. Chorianopoulos I, Tsilimigkas G, Koukoulas S, Balatsos T (2014) The shift to competitiveness and a new phase of sprawl in the Mediterranean city: enterprises guiding growth in Messoghia—Athens. Cities 39:133–143
54. Ferrara A, Salvati L, Sateriano A, Carlucci M, Gitas I, Biasi R (2016) Unraveling the 'stable' landscape: a multi-factor analysis of unchanged agricultural and forest land (1987–2007) in a rapidly-expanding urban region. Urban Ecosystems 19(2):835–848
55. Foley J, DeFries R, Asner GP, Barford C, Bonan G, Carpenter SR, Chapin FS, Coe MT, Daily GC, Gibbs HK, Helkowski JH, Holloway T, Howard EA, Kucharik CJ, Monfreda C, Patz JA, Prentice IC, Ramankutty N, Snyder PK (2005) Global consequences of land use. Science 309:570–574
56. Makhzoumi JM (1997) The changing role of rural landscapes: olive and carob multi-use tree plantations in the semiarid Mediterranean. Landsc Urban Plan 37:115–122
57. Colantoni A, Grigoriadis E, Sateriano A, Venanzoni G, Salvati L (2016) Cities as selective land predators? a lesson on urban growth, deregulated planning and sprawl containment. Sci Total Environ 545:329–339
58. Economidou E (1993) The Attic landscape throughout the centuries and its human degradation. Landsc Urban Plan 24:33–37
59. Paetzold A, Warren PH, Maltby LL (2010) A framework for assessing ecological quality based on ecosystem services. Ecol Complex 7(3):273–281
60. Raudsepp-Hearne C, Peterson GD, Bennett EM (2010) Ecosystem service bundles for analyzing tradeoffs in diverse landscapes. Proc Natl Acad Sci USA 107(11):5242–5247
61. Duvernoy I, Zambon I, Sateriano A, Salvati L (2018) Pictures from the Other Side of the Fringe: Urban growth and Peri-urban agriculture in a post-industrial city (Toulouse, France). J Rural Stud 57:25–35
62. Di Feliciantonio C, Salvati L (2015) Southern' alternatives of urban diffusion: investigating settlement characteristics and socioeconomic patterns in three Mediterranean regions. Tijdschr Econ Soc Geogr 106(4):453–470
63. Chorianopoulos I, Pagonis T, Koukoulas S, Drymoniti S (2010) Planning, competitiveness and sprawl in the Mediterranean city: the case of Athens. Cities 27:249–259
64. Cuadrado-Ciuraneta S, Durà-Guimerà A, Salvati L (2017) Not only tourism: unravelling suburbanization, second-home expansion and "rural" sprawl in Catalonia Spain. Urban Geogr 38(1):66–89
65. Ferrara A, Salvati L, Sabbi A, Colantoni A (2014) Urbanization, soil quality and rural areas: towards a spatial mismatch? Sci Total Environ 478:116–122
66. Giourga C, Loumou A (2006) Assessing the impact of pluriactivity on sustainable agriculture. a case study in rural areas of Beotia in Greece. Environ Manage 37:753–763
67. Chas-Amil ML, Touza J, García-Martínez E (2013) Forest fires in the wildland–urban interface: a spatial analysis of forest fragmentation and human impacts. Appl Geogr 43:127–137
68. Cohen M, Bilodeau C, Alexandre F, Godron M, Andrieu J, Gresillon E, Garlatti F, Morganti A (2015) What is the plant biodiversity in a cultural landscape? a comparative, multi-scale and interdisciplinary study in olive groves and vineyards (Mediterranean France). Agr Ecosyst Environ 212:175–186
69. Giannakourou G (2005) Transforming spatial planning policy in Mediterranean countries: Europeanization and domestic change. Eur Plan Stud 13:319–331
70. Amato F, Tonini M, Murgante B, Kanevski M (2018) Fuzzy definition of rural Urban interface: an application based on land use change scenarios in Portugal. Environ Model Softw 104:171–187
71. Barbati A, Corona P, Salvati L, Gasparella L (2013) Natural forest expansion into suburban countryside: Gained ground for a green infrastructure? Urban For Urban Green 12(1):36–43
72. Cimini D, Tomao A, Mattioli W, Barbati A, Corona P (2013) Assessing impact of forest cover change dynamics on high nature value farmland in Mediterranean mountain landscape. Ann SilvUltural Res 37(1):29–37

73. Karamesouti M, Detsis V, Kounalaki A, Vasiliou P, Salvati L, Kosmas C (2015) Land-use and land degradation processes affecting soil resources: evidence from a traditional Mediterranean cropland (Greece). CATENA 132:45–55

MIX
Papier aus verantwortungsvollen Quellen
Paper from responsible sources
FSC® C105338

If you have any concerns about our products,
you can contact us on
ProductSafety@springernature.com

In case Publisher is established outside the EU,
the EU authorized representative is:
Springer Nature Customer Service Center GmbH
Europaplatz 3, 69115 Heidelberg, Germany

Printed by Libri Plureos GmbH
in Hamburg, Germany